HARCOURT SCHOOL PUBLISHERS

Think Math!

Student Handbook

Developed by Education Development Center, Inc. through National Science Foundation Grant No. ESI-0099093

Visit *The Learning Site!*
www.harcourtschool.com/thinkmath

Copyright © by Education Development Center, Inc.

All rights reserved. No part of this publication may be reproduced or transmitted in any form or by any means, electronic or mechanical, including photocopy, recording, or any information storage and retrieval system, without permission in writing from the publisher.

Requests for permission to make copies of any part of the work should be addressed to School Permissions and Copyrights, Harcourt, Inc., 6277 Sea Harbor Drive, Orlando, Florida 32887-6777. Fax: 407-345-2418.

HARCOURT and the Harcourt Logo are trademarks of Harcourt, Inc., registered in the United States of America and/or other jurisdictions.

Printed in the United States of America

ISBN 13: 978-0-15-342475-5

ISBN 10: 0-15-342475-3

1 2 3 4 5 6 7 8 9 10 032 16 15 14 13 12 11 10 09 08 07

If you have received these materials as examination copies free of charge, Harcourt School Publishers retains title to the materials and they may not be resold. Resale of examination copies is strictly prohibited and is illegal.

Possession of this publication in print format does not entitle users to convert this publication, or any portion of it, into electronic format.

This program was funded in part through the National Science Foundation under Grant No. ESI-0099093. Any opinions, findings, and conclusions or recommendations expressed in this program are those of the authors and do not necessarily reflect the views of the National Science Foundation.

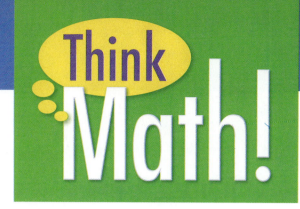

Chapter 1
Building Operations

Student Letter	1
World Almanac for Kids: *Animal Extremes*	2
Lesson 1 Explore: Strategies for Counting	4
Lesson 3 Review Model: Introducing Addition Puzzles	5
Lesson 4 Explore: Locating 6 on a Number Line	6
Lesson 6 Review Model: Using a Number Line to Add and Subtract	7
Lesson 7 Explore: Number Patterns	8
Lesson 8 Review Model: Completing Number Sentences	9
Lesson 9 Review Model: Problem Solving Strategy: *Act It Out*	10
Vocabulary	12
Game: *Number Line Grab*	14
Game: *Missing Operation Signs*	15
Challenge	16

Chapter 2
Multiplication Situations

Student Letter	17
World Almanac for Kids: *Rocks in Rows*	18
Lesson 3 Explore: Exploring Lines and Intersections	20
Lesson 3 Review Model: Intersections	21
Lesson 5 Explore: Exploring Hidden Intersections	22
Lesson 6 Explore: Exploring Pairs of Objects	23
Lesson 6 Review Model: Pairing Diagrams	24
Lesson 7 Review Model: Listing Combinations	25
Lesson 8 Review Model: Writing Multiplication Sentences	26
Lesson 10 Explore: Exploring Factors	27
Lesson 12 Review Model: Problem Solving Strategy: *Draw a Picture*	28
Vocabulary	30
Game: *Fit!*	32
Game: *Factor Maze*	33
Challenge	34

Chapter 3
Using Addition and Subtraction

Student Letter	35
World Almanac for Kids: *Tropical Fruits*	36
Lesson 1 Review Model: What is a Magic Square?	38
Lesson 2 Explore: Magic Squares with Missing Addends	39
Lesson 5 Explore: Exploring Odd and Even Numbers	40
Lesson 6 Review Model: Exchanging Coins	41
Lesson 8 Explore: Eliminating Possibilities	42
Lesson 8 Review Model: Estimating Sums and Differences	43
Lesson 9 Review Model: Problem Solving Strategy: *Work Backward*	44
Vocabulary	46
Game: *What Are My Coins?*	48
Game: *Least to Greatest*	49
Challenge	50

Chapter 4

Grouping, Regrouping, and Place Value

Student Letter	51
World Almanac for Kids: *The Grand Canyon*	52
Lesson 2 Explore: Combining and Removing Coins	54
Lesson 3 Review Model: Using Base-Ten Blocks	55
Lesson 4 Explore: Listing Possible Numbers	56
Lesson 5 Explore: Considering Digits	57
Lesson 6 Explore: Ordering Clues	58
Lesson 7 Review Model: Using a Place-Value Chart	59
Lesson 8 Review Model: Problem Solving Strategy: *Make an Organized List*	60
Vocabulary	62
Game: *Trading to 1,000*	64
Game: *Place Value Game*	65
Challenge	66

Chapter 5

Understanding Addition and Subtraction Algorithms

Student Letter	67
World Almanac for Kids: *All Kinds of Puzzles*	68
Lesson 1 Review Model: Comparing Numbers	70
Lesson 3 Explore: Jumping on the Number Line	71
Lesson 3 Review Model: Using the Number Line to Find Differences	72
Lesson 4 Explore: Predicting Digits	73
Lesson 5 Explore: Predicting Sums	74
Lesson 5 Review Model: Addition with Regrouping	75
Lesson 8 Explore: Predicting Differences	76
Lesson 9 Review Model: What Is a Cross Number Puzzle?	77
Lesson 11 Review Model: Problem Solving Strategy: *Solve a Simpler Problem*	78
Vocabulary	80
Game: *Ordering Numbers*	82
Game: *Least to Greatest*	83
Challenge	84

v

Chapter 6
Rules and Patterns

Student Letter	85
World Almanac for Kids: *Beading Fun*	86
Lesson 2 Review Model: Recording Data	88
Lesson 4 Explore: Exploring FAR Cards	89
Lesson 6 Explore: Exploring a Pattern	90
Lesson 6 Review Model: Finding a Rule	91
Lesson 7 Explore: Exploring the Number Line Hotel	92
Lesson 7 Review Model: Adding and Subtracting on a Grid	93
Lesson 8 Explore: Exploring Sharing Machine A	94
Lesson 8 Review Model: Writing Division Sentences	95
Lesson 11 Review Model: Problem Solving Strategy: *Look for a Pattern*	96
Vocabulary	98
Game: *Find a Rule*	100
Game: *Make a Rule*	101
Challenge	102

Chapter 7
Fractions

Student Letter	103
World Almanac for Kids: *Paper Folding Fun*	104
Lesson 1 Review Model: Understanding Fractions	106
Lesson 2 Review Model: Finding Equivalent Fractions	107
Lesson 4 Explore: Cracked Eggs	108
Lesson 5 Explore: Parts of a Dozen	109
Lesson 6 Explore: Fractions of an Hour	110
Lesson 6 Review Model: Using Models to Compare Fractions	111
Lesson 7 Review Model: Problem Solving Strategy: *Make a Model*	112
Vocabulary	114
Game: *Fraction Construction Zone*	116
Game: *Marble Mystery*	117
Challenge	118

Chapter 8
Charts and Graphs

Student Letter		119
World Almanac for Kids: *Wheels in the Air*		120
Lesson 3	Explore: Different Pictographs, Same Data	122
Lesson 4	Explore: Tossing Two Number Cubes	123
Lesson 4	Review Model: Describing the Likelihood of an Event	124
Lesson 5	Review Model: Listing Outcomes	125
Lesson 6	Explore: Prices at the Class Store	126
Lesson 8	Review Model: Using a Map Grid	127
Lesson 10	Review Model: Problem Solving Strategy: Make a Table	128
Vocabulary		130
Game: *Where's My House?*		132
Game: *Where's My Car?*		133
Challenge		134

Chapter 9
Exploring Multiplication

Student Letter		135
World Almanac for Kids: *Collections*		136
Lesson 1	Explore: Exploring Products	138
Lesson 1	Review Model: Using Models to Multiply	139
Lesson 2	Explore: Missing Factors	140
Lesson 2	Review Model: Fact Families	141
Lesson 5	Explore: Using 10 as a Factor	142
Lesson 6	Review Model: Making Simpler Problems	143
Lesson 7	Review Model: Problem Solving Strategy: *Guess and Check*	144
Vocabulary		146
Game: *Tic-Tac-Toe Multiplication*		148
Game: *Caught in the Middle*		149
Challenge		150

Chapter 10
Length, Area, and Volume

Student Letter	....	**151**
World Almanac for Kids: *Fancy Fish*	....	**152**
Lesson 1	Review Model: Measuring to the Nearest Inch, $\frac{1}{2}$ Inch, and $\frac{1}{4}$ Inch	**154**
Lesson 5	Explore: Measuring Area and Perimeter	**155**
Lesson 5	Review Model: Measuring Perimeter and Area	**156**
Lesson 6	Explore: An Ant Corral	**157**
Lesson 7	Explore: Building with Cubes	**158**
Lesson 7	Review Model: Measuring Volume	**159**
Lesson 8	Review Model: Problem Solving Strategy: *Draw a Picture*	**160**
Vocabulary		**162**
Game: *Ruler Game*		**164**
Game: *Perimeter Golf*		**165**
Challenge		**166**

Chapter 11
Geometry

Student Letter	....	**167**
World Almanac for Kids: *Monumental Geometry*		**168**
Lesson 2	Explore: Exploring Parallel Sides	**170**
Lesson 2	Review Model: Identifying Parallel Lines	**171**
Lesson 5	Explore: Exploring Polygons	**172**
Lesson 5	Review Model: Sorting Polygons	**173**
Lesson 8	Explore: Going on a Figure Safari	**174**
Lesson 9	Comparing Three-Dimensional Figures	**175**
Lesson 10	Review Model: Problem Solving Strategy: *Look for a Pattern*	**176**
Vocabulary		**178**
Game: *What's My Rule?*		**180**
Game: *Polygon Bingo*		**181**
Challenge		**182**

Chapter 12
Multiplication Strategies

Student Letter 183
World Almanac for Kids:
  At the Post Office 184
Lesson 1 Explore: Multiplying
           Money 186
Lesson 3 Explore: Multiplying
           Blocks 187
Lesson 4 Review Model: Using
           Models to Find Larger
           Products 188
Lesson 5 Review Model: Using
           Rectangles to Represent
           Arrays 189
Lesson 6 Explore: Separating
           Arrays 190
Lesson 8 Explore: Division
           Situation 191
Lesson 9 Review Model:
           Problem Solving Strategy:
           Work Backward 192
Vocabulary 194
Game: *Multiplication Challenge* 196
Game: *Factor Tic-Tac-Toe* 197
Challenge 198

Chapter 13
Time, Temperature, Weight, and Capacity

Student Letter 199
World Almanac for Kids: Winter
  Pleasures: *Cold and Hot* 200
Lesson 3 Explore: How Much
           Time Is Left? 202
Lesson 3 Review Model: Time,
           Distance, and Speed 203
Lesson 5 Explore: Estimating
           Weight 204
Lesson 6 Review Model: Comparing
           Capacities 205
Lesson 9 Review Model:
           Problem Solving Strategy:
           Act It Out 206
Vocabulary 208
Game: *The Freezing Game* 210
Game: *Time Concentration* 211
Challenge 212

Chapter 14
Addition and Subtraction in Depth

Student Letter 213
World Almanac for Kids:
  A Visit to New York City 214
Lesson 2 Explore: Exploring Multi-Digit Addition 216
Lesson 2 Review Model: Using Expanded Form to Add ... 217
Lesson 3 Explore: Exploring Multi-Digit Subtraction ... 218
Lesson 3 Review Model: Using Expanded Form to Subtract 219
Lesson 4 Explore: Exploring Addition and Subtraction 220
Lesson 6 Explore: Exploring Situations 221
Lesson 7 Review Model: Problem Solving Strategy: *Solve a Simpler Problem* 222
Vocabulary 224
Game: *Place Value Game* 226
Game: *Addition Scramble* 227
Challenge 228

Chapter 15
Multiplication and Division

Student Letter 229
World Almanac for Kids: *Butterflies* ... 230
Lesson 1 Explore: Dime Arrays 232
Lesson 3 Review Model: Adding Partial Products 233
Lesson 4 Explore: Multiplying with Blocks and Money 234
Lesson 6 Explore: Finding Missing Streets 235
Lesson 7 Explore: A Division Puzzle Challenge 236
Lesson 8 Review Model: Understanding Remainders 237
Lesson 9 Review Model: Problem Solving Strategy: *Draw a Picture* 238
Vocabulary 240
Game: *Factor Factory* 242
Game: *Partial Claim* 243
Challenge 244

Resources 245
Table of Measures 246
Glossary 247
Index 257

Chapter 1 Building Operations

Dear Student,

You have already had a lot of experience with counting, but have you ever thought about why it is silly to count like this?

In this chapter, you will use your experience with counting to investigate topics, such as **number lines**.

You can use your counting skills to build your skills using operations, such as **addition** and **subtraction**.

Mathematically yours,
The authors of *Think Math!*

Animal Extremes

Animals come in all shapes and sizes. Here are some of the BIGGEST animals in the world. Are any of them your favorite animal?

Tallest bird: ostrich, 9 feet tall

Longest marine mammal: blue whale, 100 feet long

Tallest mammal: giraffe, 18 feet tall

Longest fish: white shark, 45 feet long

Longest snake: python, 26–32 feet long

FACT·ACTIVITY 1

Use the animal information for Problems 1–4.

1. One blue whale is 10 feet shorter than the longest blue whale. How long is it?

2. The largest elephant can be 13 feet tall at shoulder height. How much shorter is it than the tallest giraffe?

3. One young elephant is 5 feet tall at its shoulder. How many feet must it grow to be as tall as an ostrich?

4. Draw a number line like the one below and mark the height of an ostrich and a giraffe on your number line.

0 2 4 6 8 10 12 14 16 18 20

A savannah is one of the landforms found in Africa. It is a dry and windy grassland with small plants and few trees. Giraffes are the tallest animal in the savannah. An adult giraffe can weigh as much as 3,000 pounds, and grow as tall as 18 feet.

FACT·ACTIVITY 2

For 1–4 use the number line and the animal facts to help complete the number sentences and find the answers.

1. Owls range from 5 to 28 inches tall. What is the difference in inches between the shortest and tallest owls? 28 − 5 = ■

2. How much shorter is a 26-foot python than the longest python? ■ − 26 = ■

3. Find the difference in length between the longest fish and the longest python. ■ ● ■ = ■

4. Would the tallest ostrich fit through your classroom door without bending its neck? Use a yardstick (3 feet long) to find out.
   - Is the tallest ostrich taller or shorter than your classroom door?
   - Explain how skip counting by 3s helped you.

CHAPTER PROJECT

Use the library or other sources to find and choose a marine mammal, a land mammal, a snake, a fish, and a bird. Write the name and length or height of each of your animals. You might also want to draw a picture of your animals. Then write number sentences that compare each of your animals to the longest or tallest animal of the same type on page 2.

ALMANAC Fact

A giraffe's tongue is 18 to 20 inches long and blue-black. Its feet are as large as a dinner plate!

Chapter 1
Lesson 1
EXPLORE
Strategies for Counting

Find the number of pretzels in each group. Look for shortcuts that can help you.

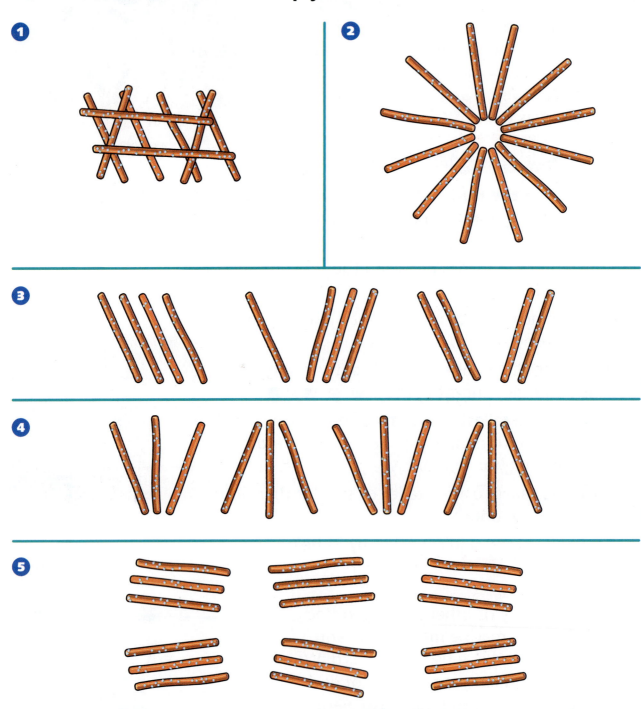

❻ Which groups have the same number of pretzels?

4 Chapter 1

Chapter 1
Lesson 3
REVIEW MODEL
Introducing Addition Puzzles

In an addition puzzle like the one shown below, the sum of the objects in two touching boxes is written in the circle between them.

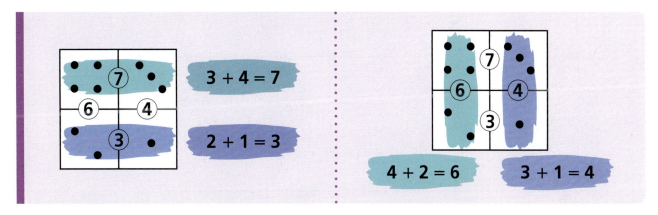

In an addition puzzle like the one shown below, the sum of the numbers in two touching boxes is written in the circle between them.

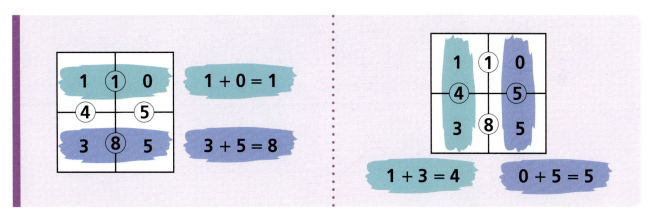

Check for Understanding

Find the missing sums.

1

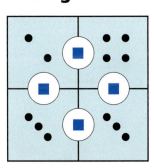

2

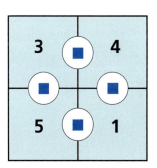

Chapter 1 **5**

Chapter 1
Lesson 4
EXPLORE
Locating 6 on a Number Line

How does the position of the number 6 make each number line different? Find the missing number for each tag.

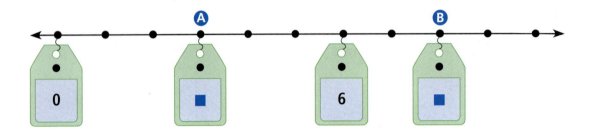

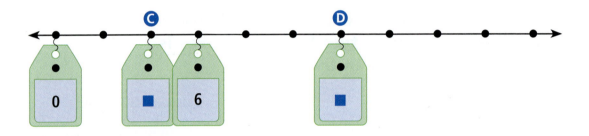

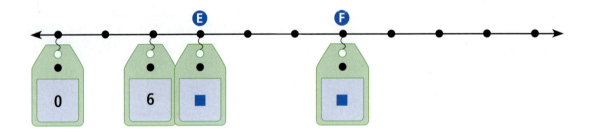

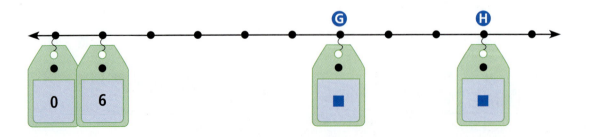

Chapter 1
Lesson 6

REVIEW MODEL
Using a Number Line to Add and Subtract

You can use a number line to add and subtract.

25 + 7 = ■

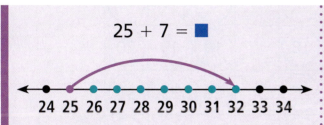

Start at 25.
Jump **forward** 7 spaces.
Land on 32.

So, 25 + 7 = 32.

40 − 6 = ■

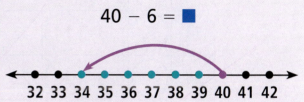

Start at 40.
Jump **backward** 6 spaces.
Land on 34.

So, 40 − 6 = 34.

You can use a number line to find a missing number.

16 + ■ = 24

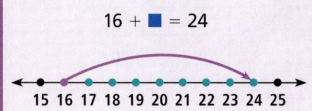

Start at 16.
Jump **forward** and land on 24.
Find the number of spaces you jumped.

So, 16 + 8 = 24.

34 − ■ = 29

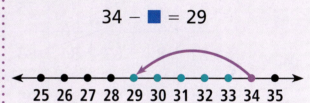

Start at 34.
Jump **backward** and land on 29.
Find the number of spaces you jumped.

So, 34 − 5 = 29.

✔ Check for Understanding

Use the number line to complete the number sentence.

33 34 35 36 37 38 39 40 41 42 43 44 45 46 47 48 49 50

1. 33 + 8 = ■
2. 46 − 9 = ■
3. 41 + ■ = 50
4. 50 − ■ = 44

Chapter 1
Lesson 7
EXPLORE
Number Patterns

Think about patterns for the marked numbers.

1	②	⊗3	④	5	⊗6	7	⑧	⊗9	⑩
11	⊗12	13	⑭	⊗15	⑯	17	⊗18	19	⑳
⊗21	㉒	23	⊗24	25	㉖	⊗27	㉘	29	⊗30
31	㉜	⊗33	㉞	35	⊗36	37	㊳	⊗39	㊵
41	⊗42	43	㊹	⊗45	㊻	47	⊗48	49	㊿
⊗51	52	53	⊗54	55	56	57	58	59	60
61	62	63	64	65	66	67	68	69	70
71	72	73	74	75	76	77	78	79	80
81	82	83	84	85	86	87	88	89	90
91	92	93	94	95	96	97	98	99	100
101	102	103	104	105	106	107	108	109	110
111	112	113	114	115	116	117	118	119	120
121	122	123	124	125	126	127	128	129	130

❶ Look at the numbers in order that are circled **blue**. If the pattern continues, what number would be circled next?

❷ If the pattern continues, what would be the next number with a **green X**?

❸ If the pattern continues, what would be the next number with both a **blue circle** and a **green X**?

8 Chapter 1

Chapter 1, Lesson 8
REVIEW MODEL
Completing Number Sentences

Number sentences contain numbers, operation signs, and an equal sign.

Some number sentences are *true*.	Some number sentences are *false*.
Examples:	Examples:
4 + 5 = 9 15 − 8 = 7	2 + 3 = 8 13 − 5 = 10

Complete the number sentence.

Paula has 14 coins in her right pocket and 6 coins in her left pocket. How many more coins does Paula have in her right pocket than in her left pocket? 14 ● 6 = 8	Which operation sign will make the sentence *true*? Try +. 14 ⊕ 6 = 8 False. Try −. 14 ⊖ 6 = 8 True. So, the correct operation sign is −.

✓ Check for Understanding

Write + or − to complete the number sentence.

1. 3 ● 5 = 8
2. 7 ● 4 = 3
3. 12 ● 8 = 4

4. 9 ● 8 = 17
5. 10 ● 5 = 15
6. 18 ● 9 = 9

7. Write a true addition sentence.

8. Write a true subtraction sentence.

Chapter 1 Lesson 9
REVIEW MODEL
Problem Solving Strategy
Act It Out

Jenny picked 6 flowers. Then she picked 7 more. Hector has 8 flowers. How many more flowers will Hector need to have the same number of flowers as Jenny?

Strategy: Act It Out

 Read to Understand

What do you know from reading the problem?

Jenny picked 6 flowers and 7 flowers. Hector has 8 flowers.

 Plan

How can you solve the problem?

You can use counters to act out the problem.

 Solve

How can you act it out?

Place 6 counters in a group, and then place 7 more counters in the same group. Place 8 counters in another group. Compare the two groups.

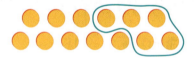 

There are 5 more counters in the first group than in second group. So, Hector will need 5 more flowers to have the same number as Jenny.

 Check

Look back at the problem. Did you answer the question that was asked? Does the answer make sense?

Problem Solving Practice

Use the strategy *act it out* to solve.

1. Andre is arranging 6 shells in 2 display boxes. How many ways can Andre separate the shells into 2 boxes?

2. There are 6 children in a group. Each child has 4 crayons. How many crayons are there in all?

Problem Solving Strategies

✓ **Act It Out**
✓ Draw a Picture
✓ Guess and Check
✓ Look for a Pattern
✓ Make a Graph
✓ Make a Model
✓ Make an Organized List
✓ Make a Table
✓ Solve a Simpler Problem
✓ Use Logical Reasoning
✓ Work Backward
✓ Write a Number Sentence

Mixed Strategy Practice

Use any strategy to solve. Explain.

3. The sum of two numbers is 20. Their difference is 4. What are the numbers?

4. Mr. Perez wrote the numbers 86, 81, 76, 71, and 66 on the board. What are the next two numbers in his pattern?

5. Lenny has 3 blue marbles, 4 red marbles, and 1 green marble in his bag. What fractional part of his bag of marbles is red?

6. Sasha has 5 stuffed dogs, 7 stuffed cats, and 3 stuffed bears. She gives 8 of the stuffed animals to her baby sister. How many stuffed animals does Sasha have left?

7. George's math group is skip-counting by fours. Each person says one number. They start with 12 and end with 36. How many people are in George's group?

8. When Megan planted the rose bush in her garden, it was 8 inches tall. Now it is 21 inches tall. How many inches has the rose bush grown?

Chapter 1 Vocabulary

Choose the best term for each sentence. Use Word List A.

1. A number is made up of at least one __?__.

2. The number 3,481 has 8 __?__.

3. A(n) __?__ has four related number sentences.

4. The operation sign is "–" for a(n) __?__.

5. The size of a jump on a number line is the number of __?__.

6. To subtract 8 – 5 on a number line, start at 8 and count 5 spaces __?__.

7. You can use a number line to __?__ by twos, threes, or fives.

Complete each analogy. Use Word List B.

8. Hundreds is to thousands as __?__ is to tens.

9. Subtraction sentence is to backward as __?__ is to forward.

Word List A
addition sentence
associative
backward
commutative
digit
fact family
forward
multiple
number line
ones
skip-count
spaces
subtraction sentence
tens
thousands

Word List B
addition sentence
commutative
digit
ones

Talk Math

Talk with a partner about what you have learned about operations. Use the vocabulary terms *number sentence*, *sum*, and *operation sign*.

10. You are given one addition sentence. How can you find the other number sentences in the fact family?

11. How can you use a number line to add?

12. How can you use a number line to subtract?

Word Definition Map

13 Create a word definition map for the term *number line*.

A What is it?

B What is it like?

C What are some examples?

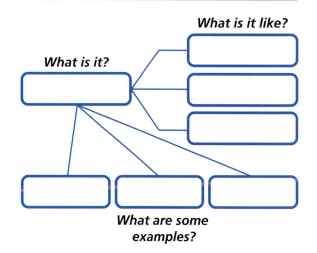

Word Line

14 Create a word line for the terms *hundreds*, *ones*, *tens*, and *thousands*.

Words:

Sequence:

SUM The word *sum* sounds just like the word *some*. Both words describe amounts. However, *sum* is an exact amount, but *some* is not. The sentence, "Ken buys *some* juice," could mean that Ken buys a glass of juice or a gallon of juice. When you need to know the *sum* of two or more numbers, you want an exact number. For example, a banker needs to know exact *sums* of money rather than *some* amount of money.

GO ONLINE Technology
Multimedia Math Glossary
www.harcourtschool.com/thinkmath

Chapter 1 **13**

GAME

Number Line Grab

Game Purpose
To practice labeling number lines, skip-counting, and adding one-digit numbers

Materials
- 3 number cubes (numbered 1–6)
- 2 different color markers, pencils, or crayons
- Activity Master 10: *Number Line Grab*

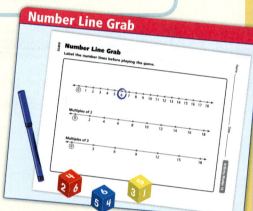

How To Play the Game

1 Work with a partner. Complete the number lines on Activity Master 10.

2 Choose a marker. Toss one of the number cubes. The player with the larger number goes first.

3 Toss all three number cubes. Choose a number to circle on any number line from these 3 choices:
- a number that matches one of the numbers tossed.
- a number that is the sum of two of the tossed numbers.
- the number that is the sum of all three tossed numbers.

Example: If you toss these numbers:

Possible numbers to circle: 1, 3, 5, 4 (1 + 3), 6 (1 + 5), 8 (3 + 5), or 9 (1 + 3 + 5)

4 Take turns tossing the number cubes and circling a number that is not already circled on any number line.
If there is no number to circle, you lose a turn.

5 The first player to circle 12 numbers is the winner.

GAME

Missing Operation Signs

Game Purpose
To practice addition and subtraction

Materials
- 2 different color markers, pencils, or crayons
- paper bag
- Activity Master 11: *Missing Operation Signs*
- Activity Master 12: *Missing Operation Signs*

How To Play The Game

1. Play with a partner. Cut out the cards from Activity Master 11. Mix them up, and choose 20. Put those cards inside a paper bag.

2. Choose a marker. Decide who will play first.

3. Without looking, take a card from the bag.
   - Find a sentence on Activity Master 12 that is missing the operation sign on your card.
   - Write the operation sign in the sentence.
   - If you cannot find a sentence for your card, you lose a turn.
   - Set aside the operation card—do not put it back in the bag.

4. Take turns taking cards from the bag and writing the operation signs in the sentences.

5. Use all the cards if you can. The game ends when there are no cards left in the bag.

6. The winner is the player who has filled in more sentences.

CHALLENGE

Robby and Ricky Rabbit like to play a jumping game on the number line. They start each game at 0 and do not go above 20.

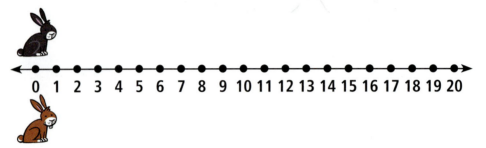

For each game, tell how many times Robby and Ricky will land on the same number. Then write all the numbers on which they will both land.

Game 1 Robby makes jumps of 3 spaces.
Ricky makes jumps of 5 spaces.

Game 2 Robby makes jumps of 2 spaces.
Ricky makes jumps of 3 spaces.

Game 3 Robby makes jumps of 3 spaces.
Ricky makes jumps of 6 spaces.

Game 4 Robby makes jumps of 2 spaces.
Ricky makes jumps of 6 spaces.

Game 5 Robby makes jumps of 6 spaces.
Ricky makes jumps of 10 spaces.

Sometimes their sister Randy Rabbit plays the jumping game with them.

Game 6 Randy makes jumps of 2 spaces. Robby makes jumps of 3 spaces. Ricky makes jumps of 4 spaces. Will all three rabbits ever land on the same number? If so, tell the number of times and the landing numbers.

Chapter 2
Multiplication Situations

Dear Student,

Do you know what situations use multiplication instead of addition? Here are some examples:

Imagine a rectangle made of dots. There are four rows of dots and three dots in each row. How many dots in all?

Imagine a tiny town. It has four streets that run north-south and three streets that run east-west. Every north-south street crosses every east-west street. How many intersections are there?

Imagine that you work in a sandwich shop. There are four kinds of sandwich fillings. There are three kinds of bread. How many "one filling-one kind of bread sandwiches" can you make?

Fillings	Bread
Tuna	White
Turkey	Wheat
Ham	Rye
Cheese	

In this chapter, you will see different ways to picture and think about multiplication as you start to learn multiplication facts.

Mathematically yours,
The authors of **Think Math!**

Rocks in Rows

FACT·ACTIVITY 1

There are three different types of rocks: igneous, sedimentary, and metamorphic. Suppose you collect rocks and have an equal number of igneous, sedimentary, and metamorphic rocks in your collection.

Use models to help you with the following problems.

1. Suppose you have 3 rocks of each type. How many rocks do you have in your collection?

2. Suppose you have 4 rocks of each type. How many rocks do you have in your collection?

3. Suppose you have 24 rocks in all. Show how you might arrange the rocks so that each row has the same number of rocks. Show four different arrays.

Ayers Rock in Australia

FACT·ACTIVITY 2

Rocks can be grouped into many different classifications including size, shape, color, and texture (how it feels). You decide to group your rocks by their color and texture.

Colors	Textures
White	Fine
Brown	Intermediate
Black	Coarse
	Glassy
	Frothy

1. You have white, brown, and black rocks. You have rocks in 5 different textures: fine, intermediate, coarse, glassy, and frothy. Make a list to show all possible combinations of color and texture for your collection of rocks.

2. Write a multiplication sentence that shows the total number of possible combinations of colors and textures.

CHAPTER PROJECT

Materials: grid paper for each student

Plan a display of 2 rock collections. One has 40 igneous rocks. The other has 60 sedimentary rocks.

- Use grid paper to show 3 possible ways to display the igneous rocks so that the same number of rocks are in each row.
- Write a multiplication sentence to represent each igneous rock display.
- Use grid paper to show 3 possible ways to display the sedimentary rocks so that the same number of rocks are in each row.
- Write a multiplication sentence to represent each sedimentary rock display.
- Use a 10 × 10 array. Show how you might arrange 40 igneous rocks and 60 sedimentary rocks together in one display.

ALMANAC Fact

Ayers Rock is one of the largest rocks in the world. It is located in central Australia, where the native (Aboriginal) people call it Uluru. It is more than 986 feet high and 5 miles around.

Chapter 2
Lesson 3
EXPLORE
Exploring Lines and Intersections

1 The map shows all the streets of a tiny town.

 A How many streets are in this town?

 B How many streets are shown as horizontal?

 C How many streets are vertical?

 D How many intersections are there?

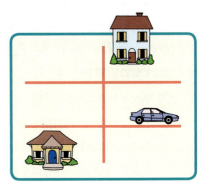

2 Draw a new map with 6 streets. All streets must be horizontal or vertical.

 A How many horizontal streets are there?

 B How many vertical streets are there?

 C How many intersections are there?

3 Draw a map with 4 streets for each problem below. Write the number intersections.

 A 0 vertical streets

 B 1 vertical street

 C 2 vertical streets

 D 3 vertical streets

 E 4 vertical streets

Chapter 2 Lesson 3 — REVIEW MODEL
Intersections

You can make a drawing to show a vertical line crossing a horizontal line. The point where the lines cross is called an intersection.

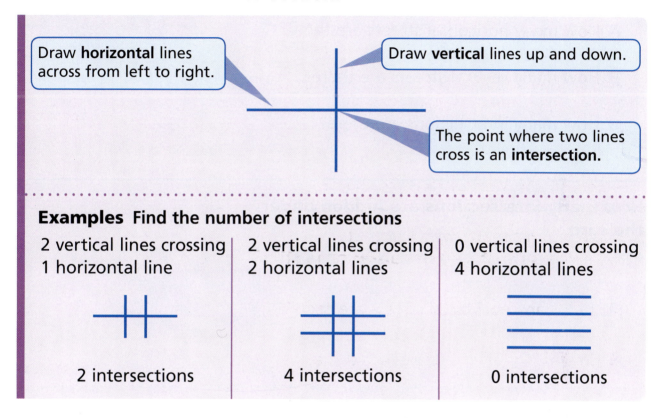

Draw **horizontal** lines across from left to right.

Draw **vertical** lines up and down.

The point where two lines cross is an **intersection**.

Examples Find the number of intersections

2 vertical lines crossing 1 horizontal line — 2 intersections

2 vertical lines crossing 2 horizontal lines — 4 intersections

0 vertical lines crossing 4 horizontal lines — 0 intersections

✓ Check for Understanding

On a separate sheet of paper make a drawing for each problem. Write the number of intersections or the number of lines.

1 3 vertical lines crossing 1 horizontal line
■ intersections

2 3 vertical lines crossing 3 horizontal lines
■ intersections

3 5 vertical lines crossing 2 horizontal lines
■ intersections

4 6 vertical lines crossing 0 horizontal lines
■ intersections

5 3 vertical lines crossing ■ horizontal lines
15 intersections

6 ■ vertical lines crossing ■ horizontal lines
12 intersections

Chapter 2 21

Chapter 2
Lesson 5
EXPLORE
Exploring Hidden Intersections

1 All the intersections are covered on this map.

   A How many horizontal streets are there?

   B How many vertical streets are there?

   C How many intersections are there?

How many intersections are hidden under the card?

2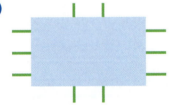

3

4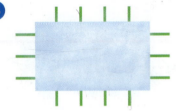

5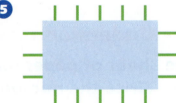

---

6 How many intersections are hidden under this card?

Chapter 2 Lesson 6
EXPLORE
Exploring Pairs of Objects

At the sandwich shop, there are 4 sandwich fillings to choose from. There are 3 kinds of bread to choose from.

Fillings	Bread
Ham	White
Peanut Butter	Wheat
Turkey	Rye
Cheese	

1 Choose one filling and one kind of bread. How many different sandwiches can you make?

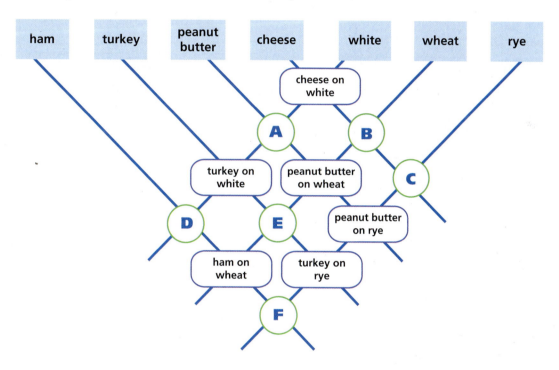

2 What sandwich names are missing from the labeled intersections?

3 The sandwich shop ran out of turkey and rye bread. How many different sandwiches can you make now?

Chapter 2, Lesson 6

REVIEW MODEL
Pairing Diagrams

You can use diagrams to show all the ways you can pair items from two different groups.

This diagram shows the pairing of ice cream flavors with topping flavors.

TODAY'S FLAVORS
Ice Cream: Strawberry, Vanilla, Chocolate, Peach
Toppings: Fudge, Butterscotch

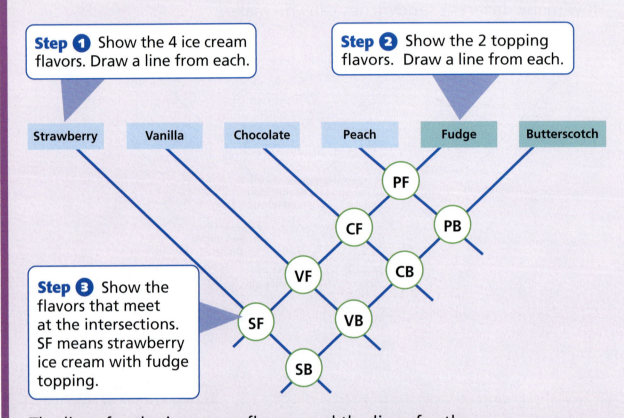

Step 1 Show the 4 ice cream flavors. Draw a line from each.

Step 2 Show the 2 topping flavors. Draw a line from each.

Step 3 Show the flavors that meet at the intersections. SF means strawberry ice cream with fudge topping.

The lines for the ice cream flavors and the lines for the topping flavors intersect at 8 points. So, there are 8 pairings.

✓ Check for Understanding

Make a drawing to show the number of pairings.

1 How many pairings of 1 kind of bread and 1 sandwich filling can you make?
Bread: white, wheat, rye
Fillings: tuna, cheese, peanut butter

2 How many pairings of 1 snack and 1 drink can you make?
Snacks: crackers, muffins, pretzels, fruit, vegetables
Drinks: juice, water, milk

Chapter 2 Lesson 7

REVIEW MODEL
Listing Combinations

You can make an organized list to show all the combinations you can make from the choices in two different groups.

This list shows all the possible combinations for one sport and one time.

Soccer in the **morning**
Soccer in the **afternoon**
Baseball in the **morning**
Baseball in the **afternoon**
Hockey in the **morning**
Hockey in the **afternoon**
Swimming in the **morning**
Swimming in the **afternoon**

Keep the list organized. One way to organize the list is to list both times for each sport in order. This will help you find all the possible combinations.

To check your work, you can multiply to find the total number of combinations.

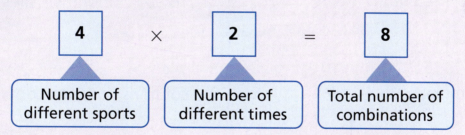

| 4 | × | 2 | = | 8 |

Number of different sports | Number of different times | Total number of combinations

The list shows 8 combinations. The multiplication sentence also shows that there should be 8 combinations possible.

✓ Check for Understanding

Make a list of all the combinations. Then write a multiplication sentence to check.

❶ **Clothes Choices**
   Pants: black, tan, green
   Shirts: red, white

❷ **Music Lesson Choices**
   Instrument: piano, guitar, drums, violin
   Days: Mon., Tue., Thu., Fri.

Chapter 2 Lesson 8

REVIEW MODEL
Writing Multiplication Sentences

You can write multiplication sentences to describe rectangular arrays.

A There are 4 rows and 3 columns.

    4 × 3 = 12
  rows columns dots

or

    3 × 4 = 12
  columns rows dots

B There is 1 row and 9 columns.

    1 × 9 = 9
  row columns dots

or

    9 × 1 = 9
  columns row dots

C There are 2 rows and 5 columns.

    2 × 5 = 10
  rows columns tiles

or

    5 × 2 = 10
  columns rows tiles

D There are 7 rows and 4 columns.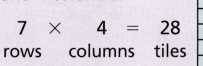

    7 × 4 = 28
  rows columns tiles

or

    4 × 7 = 28
  columns rows dots

✓ Check for Understanding

Write a multiplication sentence to describe each array.

1

2

3

Chapter 2
Lesson 10 — EXPLORE
Exploring Factors

1 **Think about maps that have 3 streets.**

A Draw all the 3-street maps.

B For each 3-street map, you can write an addition sentence like this:

■ + ■ = 3

horizontal   vertical

Write an addition sentence for each of your 3-street maps.

2 **Think about maps that have 6 intersections.**

A Draw all the 6-intersection maps.

B For each 6-intersection map, you can write an multiplication sentence like this:

■ × ■ = 6

horizontal   vertical

Write a multiplication sentence for each of your 6-intersection maps.

C Could there be a 6-intersection map with 0 horizontal streets? Draw a map or explain.

D Could there be a 6-intersection map with more than 6 horizontal streets? Draw a map or explain.

E Could there be a 6-intersection map with 4 horizontal streets? Draw a map or explain.

Chapter 2
Lesson 12

REVIEW MODEL
Problem Solving Strategy
Draw a Picture

> Carl is planning a garden. He wants to plant 48 seeds in 8 rows. How many seeds will be in each row?

Strategy: Draw a Picture

 Read to Understand

What do you know from reading the problem?

Carl is planting 48 seeds in 8 rows.

 Plan

How can you solve the problem?

You can draw a picture to find the number of seeds in each row.

 Solve

How can you draw a picture of the problem?

You can draw an array using dots to represent the seeds. Draw a column of 8 dots, one dot for each row of seeds. Add 1 dot to each row until you have drawn a total of 48 dots. If you count the number of dots in each row, you will find 6 dots. So, there will be 6 seeds in each row.

 Check

Look back at the problem. Did you answer the question that was asked? Does the answer make sense?

Problem Solving Practice

Problem Solving Strategies
- ✓ Act It Out
- ✓ **Draw a Picture**
- ✓ Guess and Check
- ✓ Look for a Pattern
- ✓ Make a Graph
- ✓ Make a Model
- ✓ Make an Organized List
- ✓ Make a Table
- ✓ Solve a Simpler Problem
- ✓ Use Logical Reasoning
- ✓ Work Backward
- ✓ Write a Number Sentence

Draw a picture to solve.

1. Jack has ten blocks numbered from 1 to 10. How many combinations of one odd-numbered block and one even-numbered block can he make?

2. The art teacher gives each student a piece of paper that is in the shape of a square. What figures can the students make if they draw a single, straight line that cuts the square in half?

Mixed Strategy Practice

Use any strategy to solve. Explain.

3. Mia wants to buy a small toy that costs 35¢. The toy is sold from a machine that only accepts quarters, dimes, and nickels. What combinations could Mia use to buy the toy?

4. Three girls stand in line at the ticket counter. Twice as many boys stand in line. How many children stand in line in all?

5. A total of 65 third and fourth graders attended this month's Math Club meeting. There were 15 more third graders than fourth graders at the meeting. How many third graders and how many fourth graders attended the meeting?

6. Daryl is ordering pizza. His topping choices are onions, extra cheese, or peppers. He can choose either a thin crust or a thick crust. How many choices does Daryl have?

Chapter 2 Vocabulary

Choose the best vocabulary term from Word List A for each sentence.

Word List A
array
column
east
grid
horizontal
intersection
north
row
separation
south
times
vertical
west

1. The directions on a map are north, south, __?__ and west.

2. A(n) __?__ line is one that goes from top to bottom.

3. A(n) __?__ has both horizontal and vertical lines in it.

4. A(n) __?__ is where two lines cross.

5. When you read 3 × 8, you say "three __?__ eight."

6. Dots arranged in columns and rows form a(n) __?__.

Complete each analogy using the best term from Word List B.

Word List B
factor
multiple
product

7. Addend is to addition as __?__ is to multiplication.

8. Sum is to addition as __?__ is to multiplication.

Talk Math

Discuss with a partner what you have learned about modeling multiplication. Use the vocabulary terms *factor*, *multiple*, and *product*.

9. How can you use an array to model multiplication?

10. How can you use intersecting lines to model multiplication?

11. How can you use multiplication to check that you have listed all possible combinations?

Tree Diagram

12 Create a tree diagram for the term *multiply*. Use the terms *column, row, array, intersection, grid, horizontal,* and *vertical.*

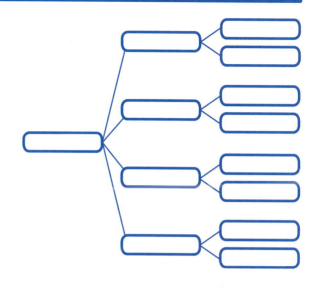

Concept Map

13 Create a concept map for *Multiplication Situations*. Use the terms *combinations, array,* and *intersections*. Describe what each represents.

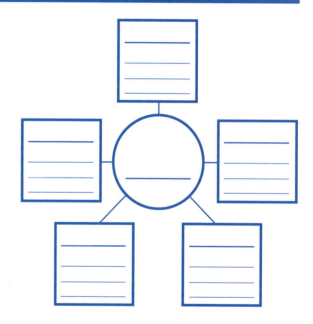

What's in a Word?

COMBINATION The word *combination* is used in different ways. The *combination* for a lock is often made up of numbers in a particular order. Suppose the *combination* to a lock is 1-2-3. The lock will not open if you try 2-3-1 or 1-3-2. In math, the order of the items does not matter. For example, blue shirt with black pants is the same *combination* as black pants with blue shirt.

GO ONLINE Technology
Multimedia Math Glossary
www.harcourtschool.com/thinkmath

Chapter 2 **31**

GAME

FIT!

Game Purpose
To form rectangular arrays

Materials
- crayons or markers
- 2 number cubes (numbered 1–6)
- about 40 beans or other objects
- Activity Master 19: *Fit!* gameboard

How to Play the Game

1 Two people can play the game. Each person chooses a different color of marker or crayon. Toss one of the number cubes. The player with the larger number goes first.

2 Start by tossing both number cubes. Use the numbers to create an array on the *Fit!* gameboard:
- Count the number of rows using one number. Then count the number of columns using the other number.
- Use the beans to make an array. Color the array.
- Label your array with the total number of tiles.
- If you can't draw an array, check off a strike box.

3 Take turns until both players have three strikes, or the gameboard is full.
- If one player gets three strikes, the other play may continue until they get three strikes or fill the board.

4 The player with more arrays is the winner.

GAME

Factor Maze

Game Purpose
To identify products with factors from 1 to 6

Materials
- number cube
- 2 different colors of crayons or markers
- Activity Masters 23 and 24

How to Play the Game

1. Toss the number cube. The player with the larger number goes first.

2. Start by tossing the number cube. Write the **Toss** number on the *Factor Maze* Recording Sheet.

3. Draw your move on the *Factor Maze* gameboard.
   - You may move one square horizontally or vertically (but not diagonally) to any square that your number is a factor of.

   For example, if you toss a 4, you can move to any square with a product that has 4 as a factor, such as 4, 8, 12, 16, 20, 24, 28, 32, 36, and so on. But you may not move to a space marked 2, because 4 is not a factor of 2.
   - Write your **Moved To** number on the *Factor Maze* Recording Sheet.

4. If the toss is not a factor of any of the numbers in adjacent squares, write an X in one of the strike boxes at the bottom of the gameboard.

5. The first player to reach the Finish square is the winner, or the first player with three strikes loses.

Chapter 2 33

CHALLENGE

Each circle has all the numbers you need to make several multiplication sentences. Can you find all the multiplication sentences for each circle? You must use all the numbers in the circle. You can use each number only once.

Good luck!

1

```
    8   3
  1   4   16
    6   6
    2   12
```

One multiplication sentence in this circle is **2 × 8 = 16**. Can you find two others?

2

```
    9   2
  9   24   3
    18   3
    3   8
```

Find 3 multiplication sentences in this circle.

3

```
    8  4  8
  6   24  12
  48   8   32
    2  8  64
```

Find 4 multiplication sentences in this circle.

4

```
    7   8   9
  21   3   63
  7   35   7
  56   5   7
```

How many multiplication sentences can you find in this circle?

34   Chapter 2

Chapter 3
Using Addition and Subtraction

Dear Student,

In Chinese legend, a curious pattern of numbers was once found on the back of a turtle shell. People have been studying the mystery of these numbers for centuries.

This arrangement of numbers has certain properties, so people began to call it a magic square. Do you know what makes the pattern of numbers in the magic square special? Could you make your own magic square?

In this chapter, you will be using your addition and subtraction skills to solve interesting questions and puzzles like the magic square. You will work with money, and you'll also learn about reading graphs similar to the graphs scientists use to see patterns.

Mathematically yours,
The authors of *Think Math!*

Tropical Fruits

Most of us are familiar with at least two types of tropical fruit: pineapples and bananas. However, there are many other fruits that grow in tropical regions.

FACT·ACTIVITY 1

The durian is known as the "King of Fruits" in parts of Southeast Asia. It is a very popular tropical fruit found mainly in Malaysia, Indonesia, and Thailand. Durians are either round or oval in shape and covered with hard spikes. Some people love the melons, but others don't because the fruit has a very strong smell.

Answer the following questions. Use the data in the Almanac Fact for Problems 2 and 3.

1. If you buy three durians weighing 5 pounds, 10 pounds, and 8 pounds, what is the total weight of the melons?

2. Suppose a young durian tree bears only 10–50 melons. How old could the durian tree be?

3. If a durian tree is 11 years old, how many more years will the tree need to grow before it can produce more than 100 melons?

A tropical fruit salad is a great way to enjoy a variety of fruits. Three different mixed tropical fruit salads are shown below. To set the price of each salad, the chef first sets a price for each kind of fruit used in the salads. The price of the salad is the sum of the prices for the fruits used in the salad.

Answer the questions based on the prices given.

Fruit Salad 1: Watermelon, Star Fruit, Pineapple

Fruit Salad 2: Mango, Papaya, Pineapple

Fruit Salad 3: Jambu, Guava, Watermelon, Pineapple

Price of Tropical Fruits per Serving	
Tropical Fruit	Price
Jambu (water apple)	3¢
Pineapple	5¢
Star Fruit	10¢
Guava	15¢
Watermelon	20¢
Papaya	25¢
Mango	30¢
Strawberries	35¢
Kiwi	40¢

1. What is the price of Fruit Salad 1?
2. If you pay for Fruit Salad 2 with 3 quarters, how much change should you receive? Explain.
3. What is the fewest number of coins you could use to buy Fruit Salad 3? Explain.

CHAPTER PROJECT

Materials: paper, markers or colored pencils

Create a menu with 3 different tropical fruit salads. Use the list of fruits and prices from Fact Activity 2. You may use no more than 5 fruits in your salad. Each fruit salad must cost a total of 99 cents or less.

- Name your salads.
- List the ingredients of your salad.
- Calculate the total price of each salad on your menu.
- List the fewest coins needed to equal the price of the salad.

ALMANAC Fact

It takes about 6 years for a durian tree to bear fruit. Trees older than 10 years can bear 50-100 melons a year. Twenty-year-old trees can bear 100-200 melons.

Chapter 3
Lesson 1
REVIEW MODEL
What Is a Magic Square?

In a magic square like this one, every row, column, and diagonal has the same sum.

7	17	3
5	9	13
15	1	11

7 + 17 + 3 = 27

5 + 9 + 13 = 27

15 + 1 + 11 = 27

7 + 5 + 15 = 27

17 + 9 + 1 = 27

3 + 13 + 11 = 27

7 + 9 + 11 = 27
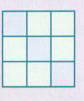

15 + 9 + 3 = 27

✓ Check for Understanding

❶ If one row has a different sum than the others, is the array a magic square? Explain.

❷ If you changed one of the numbers in this magic square to an even number, would it still be a magic square? Explain.

Chapter 3 Lesson 2

EXPLORE
Magic Squares with Missing Addends

Fill in the blanks so that each row, column, and diagonal in the magic square has the same sum.

5	■	11
■	14	■
■	■	23

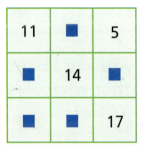

① What is the special sum for this magic square?
How do you know?

② What is the special sum for this magic square?
How do you know?

③ Describe any relationships you see between the two completed magic squares.

Chapter 3 • 39

Chapter 3
Lesson 5

EXPLORE
Exploring Odd and Even Numbers

> Jackie had some marbles and two boxes—one blue box and one green box. Jackie put all the marbles in the two boxes.

What can you say about this situation?
Write whether each statement is *true* or *false*.

Statement ❶

If Jackie started with an even number of marbles, then she **must** have put the same number in each box.

Statement ❷

If Jackie started with an even number of marbles, then she **could** have put the same number in each box.

Statement ❸

If Jackie started with an odd number of marbles, then she **could not** have put the same number in each box.

Statement ❹

If Jackie put the same number of marbles in each box, then the total number of marbles **must** have been even.

Statement ❺

If Jackie put a different number of marbles in each box, then the total number of marbles **must** have been odd.

Chapter 3 Lesson 6
REVIEW MODEL
Exchanging Coins

You can practice regrouping by exchanging coins to find the fewest coins for a given amount.

Activity Use quarters, dimes, nickels, and pennies to make 72¢. When possible, exchange for coins of greater value to find the fewest coins for 72¢.

Step 1
Look at the collection of coins. Is the amount made with the fewest coins?

The amount is not made with the fewest coins, because you can exchange some pennies for another nickel.

Step 2
Exchange 5 pennies for 1 nickel.

Can another exchange be made?

Step 3
Exchange 2 nickels for 1 dime.

No more exchanges can be made. So, the collection has the fewest coins possible for 72¢.

✓ Check for Understanding

Show how to make the same amount using the fewest coins. Use quarters, dimes, nickels, and pennies.

1

2

3

Chapter 3 Lesson 8
EXPLORE
Eliminating Possibilities

When taking tests, you can use estimation to eliminate wrong answer choices.

Problem	Without calculating, quickly choose one answer that you are sure is wrong. Explain what makes you sure.	Without looking at any of the possible answers, make a quick estimate.	Choose the answer.
❶ 56 +47 A. 9 B. 32 C. 93 D. 103		Estimate:	A. 9 B. 32 C. 93 D. 103
❷ 47 −38 A. 9 B. 19 C. 75 D. 85		Estimate:	A. 9 B. 19 C. 75 D. 85
❸ 83 −48 A. 21 B. 35 C. 45 D. 131		Estimate:	A. 21 B. 35 C. 45 D. 131
❹ 736 +264 A. 532 B. 900 C. 990 D. 1,000		Estimate:	A. 532 B. 900 C. 990 D. 1,000

Hint . . . You may want to write something like, "about 1,000," or "between 900 and 1,100."

42 Chapter 3

Chapter 3 Lesson 8
REVIEW MODEL
Estimating Sums and Differences

You can estimate sums and differences by using various methods.

Estimate 32 + 29.

One Way

Step 1
Use the digit in the greatest place-value position to approximate each number.

32 + 29
30 + 20 = 50

Step 2
Get a closer estimate by seeing if the digit in the next greatest place-value position will have an effect.

32 + 29
2 + 9 will make another 10.
30 + 20 = 50 and 50 + 10 = 60

So, the sum is about 60.

Another Way

Use the closest multiples of 10.

32 + 29

The closest multiple of 10 for 32 is 30.

The closest multiple of 10 for 29 is 30.

30 + 30 = 60

So, the sum is about 60.

✓ Check for Understanding

Estimate the sum or difference. Choose any method.

1.   43
    + 28

2.   84
    − 66

3.   602
    + 275

Chapter 3 Lesson 9

REVIEW MODEL
Problem Solving Strategy
Work Backward

> Antonio has some nickels in his pocket. He puts 4 dimes in his pocket. Now he has 65¢. How many nickels does Antonio have?

Strategy: Work Backward

Read to Understand

What do you know from reading the problem?

When Antonio put 4 dimes in his pocket he had 65¢.

What do you need to find out?

The number of nickels Antonio has.

Plan

How can you solve the problem?

You can work backward.

Solve

How can you work backward to solve the problem?

Start with the total amount Antonio has: 65¢. Subtract 40¢ for the 4 dimes he put in his pocket: 65¢ − 40¢ = 25¢. The difference of 25¢ represents the value of the nickels in his pocket. So, Antonio has 5 nickels.

Check

Look back at the problem. Did you answer the question that was asked? Does the answer make sense?

Problem Solving Practice

Work backward to solve.

1. Ty has 18 marbles. Wes has 12 more marbles than Zach. Zach has the same number of marbles as Ty. How many marbles does Wes have?

2. Before going to the 2:45 P.M. movie, you need to practice the piano for 1 hour and then clean your room for 15 minutes. Allow 30 minutes to walk to the theater. What is the latest time you could begin piano practice and still get to the movie on time?

Problem Solving Strategies

- ✔ Act It Out
- ✔ Draw a Picture
- ✔ Guess and Check
- ✔ Look for a Pattern
- ✔ Make a Graph
- ✔ Make a Model
- ✔ Make an Organized List
- ✔ Make a Table
- ✔ Solve a Simpler Problem
- ✔ Use Logical Reasoning
- ✔ **Work Backward**
- ✔ Write a Number Sentence

Mixed Strategy Practice

Use any strategy to solve. Explain.

3. On the first day of your 5-day vacation, you collect 25 shells. You collect 20 on the second day, 16 on the third day, and 13 on the fourth day. If the pattern continues, how many shells will you collect on the fifth day?

4. Today's lunch menu lists turkey and roast beef. The vegetable choices are carrots, green beans, or broccoli. How many different choices of one meat and one vegetable are there?

5. Ice cream costs 50¢. Jackson has 2 quarters, 2 dimes, and 3 nickels. Find all the ways he can use his coins to pay for the ice cream with the exact amount.

6. Jennifer bought a cap and shirt and spent $27. The shirt cost twice as much as the cap. What was the cost of each item?

Chapter 3   Vocabulary

Choose the best vocabulary term from Word List A for each sentence.

1. A(n) __?__ describes a number that is close to an exact amount.
2. A horizontal line on a grid makes a(n) __?__ with a vertical line.
3. You get two coins when you __?__ a dime for nickels.
4. There are 4 different coin __?__ that make 11¢.
5. The __?__ of two dimes is 20¢.
6. If you __?__ do something, then you are required to do it.
7. A(n) __?__ number of marbles can be divided equally between two girls.

Word List A

addition sentence
amount
arrangement
array
combinations
could
estimate
even
exchange
intersection
must
odd
reflection
regroup
subtraction sentence
value

Complete each analogy. Use the best term from Word List B.

8. Greatest is to most as __?__ is to least.
9. Sum is to addition as __?__ is to subtraction.

Word List B

difference
fewest
greatest
most
sum

Talk Math

Discuss with a partner what you have learned about magic squares. Use the vocabulary terms *sum*, *column*, *row*, and *diagonal*.

10. How can you describe a magic square?
11. How can you prove that a magic square is really a magic square?

46   Chapter 3

Word Web

12. **Create a word web for the term *reflection*. Use what you have learned about reflections of magic squares.**

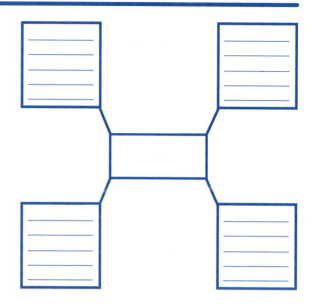

Word Definition Map

13. **Create a word definition map for the term *intersection*.**

    A What is it?

    B What is it like?

    C What are some examples?

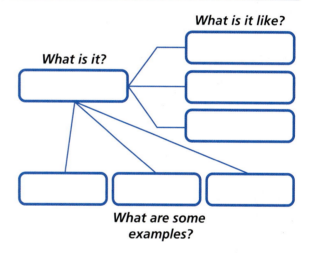

What's in a Word?

MAGIC SQUARE Magic squares have been around for at least 3,000 years. The Chinese are the first people known to have made them. Not all magic squares have 3 rows and 3 columns. Benjamin Franklin was a famous American who loved magic squares. When he was a boy, he made one with 8 rows and 8 columns. Then a friend showed him a magic square with 16 rows and 16 columns. In a magic square, the number of columns must be the same as the number of rows. Also, the sum of each row, column, and diagonal must be equal.

GO ONLINE Technology
Multimedia Math Glossary
www.harcourtschool.com/thinkmath

Chapter 3   47

GAME

What Are My Coins?

Game Purpose
To practice finding the value of a group of coins

Materials
- A collection of real or play coins: pennies, nickels, dimes, and quarters

How to Play the Game

1 Play this game with a partner. Choose about 10 coins. Place them on the table between you. Then toss a coin to decide who plays first. The first player is the Hider. The second player is the Guesser.

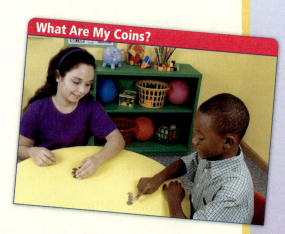

2 While the Guesser looks away, the Hider takes some of the coins, finds their value, and hides them.

3 The Guesser asks *yes or no* questions to figure out which coins are hidden.

Examples:
- Are any of the coins pennies?
- Are the coins worth more than 20¢?
- Do you have more than three different coins?

4 When the Guesser has named the hidden coins and their total value, players switch roles. Put the hidden coins back on the table and play again.

Least to Greatest

Game Purpose
To practice estimating sums and differences

Materials
- Activity Masters 33 and 34: *Least to Greatest Cards*
- Stopwatch or clock with a second hand

How to Play the Game

1 Cut out the Least to Greatest cards from Activity Masters 33 and 34. Mix up the cards. Put them in a pile face down.

2 Play this game with a partner. One player is the Placer. The other player is the Timer. The Timer times the Placer for 60 seconds with a stopwatch or the second hand on a clock.

3 The Timer says "Go." The Placer turns over the cards one at a time. The goal is to place as many cards as possible in order from the smallest sum or difference to the largest. The Placer can pass on any card by setting it aside.

4 The Timer says "Stop" at 60 seconds and checks the order of the cards. The Timer tells the Placer if there are any mistakes. The Timer does not say what the mistakes are.

5 The Placer can try to correct the order of the cards. The Placer can even remove cards. When both players agree that the order is correct, the Placer gets 1 point for each card.

Example: These four cards are placed correctly.

| 19 − 13 | 46 − 29 | 47 − 23 | 16 + 17 |

6 Switch roles. Play until time is called. The player with more points wins the game.

CHALLENGE

You can start with one magic square and change it to another. Complete the first problem below. Use that result to help you look for a pattern in the other magic squares. Can you predict the sum for each new magic square?

❶ Find the sum of this magic square.

15	18	3
0	12	24
21	6	9

For a new magic square, add 2 to each number.

A I predict that the sum of the new magic square will be ■.

B Draw the new magic square. What is the sum?

❷ Find the sum of this magic square.

17	32	29
38	26	14
23	20	35

For a new magic square, subtract 4 from each number.

A I predict that the sum of the new magic square will be ■.

B Draw the new magic square. What is the sum?

❸ Find the sum of this magic square.

9	2	7
4	6	8
5	10	3

For a new magic square, multiply each number by 2.

A I predict that the sum of the new magic square will be ■.

B Draw the new magic square. What is the sum?

Chapter 4
Grouping, Regrouping, and Place Value

Dear Student,

Why is it that 2 dimes and 4 pennies are 24 pennies, but **2 feet** and **4 inches** are not **24 inches**? You will explore questions like this one as you discuss different ways to group objects. These measurement units count things in different ways, so the amounts are written in different ways.

What about other measurement units you know?

How many days in **1 week**?
How many days in **2 weeks**?

How many minutes in **1 hour**?
How many minutes in **3 hours**?

Mathematically yours,
The authors of **Think Math!**

The Grand Canyon

FACT·ACTIVITY 1

The Grand Canyon in northern Arizona was formed over millions of years as the Colorado River eroded away the land to make a deep gorge. Although the canyon is in a very dry area, many different kinds of wildlife live there.

For these problems, use base-ten blocks where a unit cube represents 1, a rod is 10, and a flat is 100.

Grand Canyon Wildlife	
Animal Life	Number of Species
Mammals	91
Reptiles and Amphibians	57
Fish	17
Birds	373

1. Which base-ten blocks could you use to show the number of reptiles and amphibians?
2. How many flats are needed to show the number of bird species?
3. How could you use base-ten blocks to help you find the total number of mammals and fish?
4. Find the sum of the fish and bird species.

FACT ACTIVITY 2

The Grand Canyon area was established as a National Park in 1919. Millions of people visit the park every year. Visitors can hike trails, take mule trips, and camp in family campgrounds.

Write the number being described. Then write the Grand Canyon fact on this page that it represents.

1. My tens digit is 3. I tell a number of miles.
2. I am a number greater than 50 that has a ones digit that is 5 less than my tens digit.
3. I am greater than the number of miles of roads and less than the number of miles of trails.
4. My word name has "two hundred" in it, but I am greater than 300.
5. If you add 200 to me and increase my ones digit by 1, you will get another number in the facts. Name both numbers and facts.

630 miles of trails

mule trip riders in one year, South Rim, Plateau Point: 5,228

284 miles of roads

Family Campsites

Name of Campground	Number of Campsites
Mather	314
Desert View	50
North Rim	83
Tuweep	12

CHAPTER PROJECT

On a trip to the Grand Canyon, you buy a postcard for $1.00 to send to your friend. You use only coins for your purchase. You use at least one of each coin (quarters, dimes, nickels, and pennies) to purchase the card.

- Make a table to show what coins you could use to purchase the postcard. Include at least 4 combinations.
- Which of your combinations uses the least number of coins?

Grand Canyon National Park

ALMANAC Fact

On the average, the walls of the Grand Canyon rise about 1 mile above the Colorado River.

Chapter 4
Lesson 2

EXPLORE
Combining and Removing Coins

1 Ari found 2 dimes and 5 pennies under the couch. Then he found 1 dime and 8 pennies in his pocket.

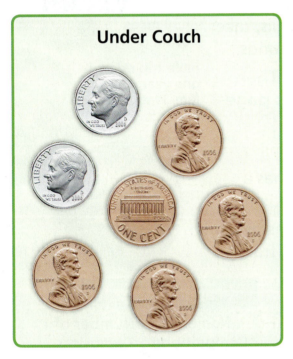
Under Couch

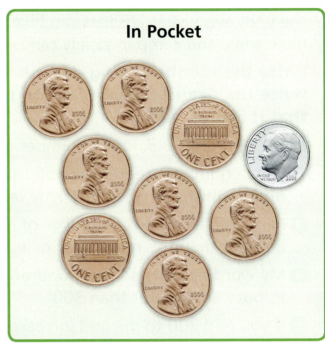
In Pocket

His mother traded coins with him so he had the same amount in fewer coins, but still had only dimes and pennies. How many dimes and pennies did he have?

---

2 Esta had 4 dimes and 3 pennies.

She bought something that cost 2 dimes and 6 pennies. If she has only dimes and pennies and the fewest coins, how much does she have left?

Chapter 4 Lesson 3
REVIEW MODEL
Using Base-Ten Blocks

You can use base-ten blocks to represent a number.

Example A

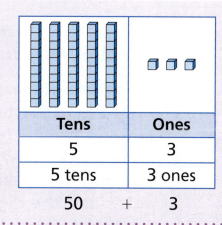

Tens	Ones
5	3
5 tens	3 ones

50 + 3 = 53

Example B

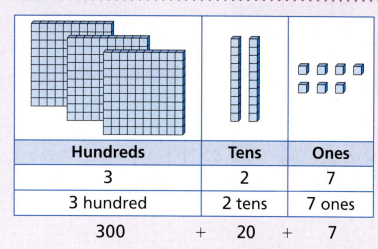

Hundreds	Tens	Ones
3	2	7
3 hundred	2 tens	7 ones

300 + 20 + 7 = 327

✓ Check for Understanding

Write the number.

1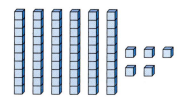

2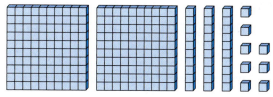

3 Use pictures, numbers, or words to tell how you would represent 419 using base-ten blocks.

Chapter 4 55

Chapter 4
Lesson 4
EXPLORE
Listing Possible Numbers

> **Who am I?**
>
> **Clue A:** I am greater than 4 X 11 and less than 5 X 11.
>
> **Clue B:** $u < t$
>
> **Clue C:** I am odd.
>
> **Clue D:** A group of 6 base-ten blocks matches me.

Mrs. Jackson loved to invent puzzles for her class.

One student guessed that the *t* stood for tens. Mrs. Jackson used *u* to stand for units, because she thought *o* (for ones) looked too much like the digit 0.

1 Make a list of the numbers that match **Clue A**.

2 A student noticed that **Clue B** eliminates 45. What numbers can you cross off the list because they do not fit **Clue B?**

3 What numbers are still on the list?

4 What numbers can you cross off the list because they do not fit **Clue C?**

5 What numbers are still on the list?

6 What is the mystery number?

Chapter 4
Lesson 5
EXPLORE
Considering Digits

The class looked at the first clue in this puzzle. Someone said there were too many numbers to list. So, the class decided to list the possible units and tens digits, and then cross out the digits that did not fit the clues.

A student wrote this list on the board:

Who Am I?
Clue A: I am odd.
Clue B: I can be made with 13 base-ten blocks.
Clue C: t > u

t	u
	0
1	1
2	2
3	3
4	4
5	5
6	6
7	7
8	8
9	9

❶ Why didn't the student write a 0 in the tens column?

❷ What is the mystery number?

Chapter 4
Lesson 6
EXPLORE
Ordering Clues

Does the order of clues change how you solve a puzzle?

1 To find the mystery number, use the clues in order.

> **Who Am I?**
> Clue A: I have 2 digits.
> Clue B: All my digits are odd.
> Clue C: The product of my digits is 5.
> Clue D: $t < u$

2 Now try using the clues in this order. How is your reasoning different?

> **Who Am I?**
> Clue A: I have 2 digits.
> Clue B: The product of my digits is 5.
> Clue C: All my digits are odd.
> Clue D: $t < u$

Chapter 4 Lesson 7
REVIEW MODEL
Using a Place-Value Chart

You can use a place-value chart to help you understand each digit in a number.

Example A

There were 29,460 people at the candidate's speech.

Ten Thousands	Thousands	Hundreds	Tens	Ones
2	9	4	6	0

Find the value of the digit 9: 9,000.
Name the number using words: twenty-nine thousand, four hundred sixty.

Example B

The candidate won the election by 852,641 votes.

Hundred Thousands	Ten Thousands	Thousands	Hundreds	Tens	Ones
8	5	2	6	4	1

Find the value of the digit 5: 50,000.
Name the number using words: eight hundred fifty-two thousand, six hundred forty-one.

✓ Check for Understanding

Write the value of the blue digit.

❶ 9,486   ❷ 309,421   ❸ 418,237

Write the number.

❹ five thousand, eight hundred forty

❺ sixty thousand

❻ two hundred thirty-one thousand, seven hundred fifty-six

Chapter 4, Lesson 8

REVIEW MODEL
Problem Solving Strategy
Make an Organized List

How many ways can you arrange the digits 4, 5, and 6 to make a three-digit number?

Strategy: Make an Organized List

Read to Understand

What do you know from reading the problem?

I need to find all the ways to arrange the digits 4, 5, and 6 to make a three-digit number.

Plan

How can you solve the problem?

You can make an organized list.

Solve

How can you make an organized list?

List the three-digit numbers with a 4 in the hundreds place and 5 or 6 in the tens and ones places.

Then list the numbers with a 5 in the hundreds place and 4 or 6 in the tens and ones places.

Finally, list the numbers with a 6 in the hundreds place and 4 or 5 in the tens and ones places.

The list shows 4, 5, and 6 can be arranged in 6 different ways.

H	T	O
4	5	6
4	6	5
5	4	6
5	6	4
6	4	5
6	5	4

Check

Look back at the problem. Did you answer the question that was asked? Does the answer make sense?

Problem Solving Practice

Use the strategy *make an organized list* to solve.

1. What is the two-digit mystery number?

   A. I am greater than 9 × 9.
   B. I am an odd number.
   C. My tens digit is 2 more than my ones digit.

2. Simon and Lily used the spinner shown at right. They spun the pointer and recorded their results. Their results were 6, 5, 6, 4, 1, 2, 1, 3, 5, 6, 2, 3, 6, 4, and 2. Which number occurred most often?

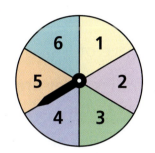

Problem Solving Strategies

✔ Act It Out
✔ Draw a Picture
✔ Guess and Check
✔ Look for a Pattern
✔ Make a Graph
✔ Make a Model
✔ **Make an Organized List**
✔ Make a Table
✔ Solve a Simpler Problem
✔ Use Logical Reasoning
✔ Work Backward
✔ Write a Number Sentence

Mixed Strategy Practice

Use any strategy to solve. Explain.

3. Javier and Nina are playing a game. Javier has 8 cards and picks up 5 cards. Nina has 6 cards. How many more cards does Nina need to have the same number as Javier?

4. Janelle spent 2 weeks and 1 day at camp. She spent 1 week and 5 days visiting her grandmother. How many days was Janelle away?

5. Jim hiked the first 3 miles of the trail in 1 hour. If he continues at the same pace, how many miles will he hike in 4 hours?

6. Olga is choosing a writing tool and a paper color for her journal. She can choose a pencil, a pen, a crayon, or a marker. She can choose white or yellow paper. What are all the different combinations of a writing tool and paper Olga can choose?

Chapter 4   61

Chapter 4 Vocabulary

Choose the best vocabulary term from Word List A for each sentence.

1. Suppose you combine groups of tens and ones. You can __?__ them to find the fewest units.

2. The place between the thousands place and the tens place is the __?__ place.

3. The __?__ in the tens place of 947 is 4.

4. You have three piles of coins. They are 6 nickels, 8 dimes, and 9 pennies. The pile with the __?__ is 6 nickels.

5. To solve a mystery number puzzle, make a list and __?__ possibilities using the clues.

6. When you __?__ 8,925 to the nearest thousand, you get 9,000.

Word List A
- digit
- eliminate
- fewest units
- hundreds
- millions
- number
- regroup
- round
- smallest units
- ten thousand
- tens
- thousands

Complete each analogy. Use the best term from Word List B.

7. Cent is to dollar as __?__ is to hundreds.

8. Letter is to word as __?__ is to number.

Word List B
- digit
- ones
- trade
- unit

Talk Math

Discuss with a partner what you have learned about place value. Use the vocabulary terms *digit*, *thousands*, *hundreds*, *tens*, and *ones*.

9. How can you round a number to the nearest thousands place?

10. How can you use base-ten blocks to represent a four-digit number?

Analysis Chart

11. Create an analysis chart for the place-value terms *hundreds, millions, ten thousands,* and *thousands.*

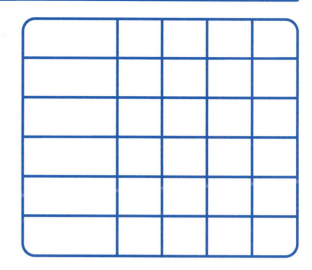

Word Web

12. Create a word web using the word *round.* Use what you know about the different meanings of *round.*

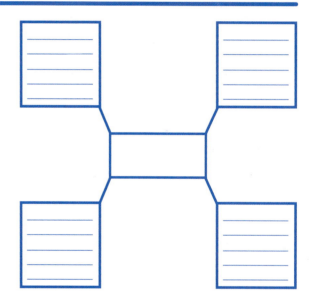

MILLION The English word *million* comes from the old Italian word *millione.* It was first used in the 1300s. *Milla* means "thousand." The suffix "*-one*" means "great." So, *millione* means "a great thousand."

Suppose the word *million* had not been created. We would have to call the millions place the *thousand-thousands place.* Then the ten-millions place would become the *ten-thousand-thousands place.* Number names would be very long and too confusing to use.

GO ONLINE Technology
Multimedia Math Glossary
www.harcourtschool.com/thinkmath

Chapter 4 **63**

GAME

Trading to 1,000

Game Purpose
To use base-ten blocks to represent sums

Materials
- 2 number cubes labeled 1–6
- Base-ten blocks (units, rods, flats, 1 large cube)
- Activity Master 35: Trading to 1,000

How To Play The Game

1. Play this game with a partner. Each player will need Activity Master 35. Decide who will play first.

2. The first player tosses both number cubes.
   - Write the numbers under Toss A and Toss B on Activity Master 35.
   - Use the tossed numbers to make a two-digit number. Write it under Chosen Number in the table.
   - Show your chosen number with base-ten blocks. Combine them with the blocks from your previous total. (There is nothing to combine on your first toss.)
   - Write an addition sentence for the combined base-ten blocks in the last column.

3. Players take turns. The first player to trade 10 flats for the large cube wins!

GAME

Place Value Game

Game Purpose
To practice identifying place-value attributes

Materials
- Activity Masters 36–46: Attribute Cards, Sets A–C
- scissors

How to Play the Game

1. Play this game with a small group. Cut out the Attribute Cards. There are three sets of cards. Choose the set (or sets) you want to use.
   - Set A cards have the easiest clues.
   - Set B cards have more difficult clues.
   - Set C cards have the most difficult clues.

2. Each player writes 5 four-digit numbers on a sheet of paper. Write neatly and large enough for others to see.

3. Place the Attribute Cards face down. Take turns turning over an Attribute Card and reading it aloud. All players cross out any of their numbers that match that attribute.

4. Play until someone crosses out all 5 numbers. That person wins!

   Example: A player turns over this card:

   Your numbers are 1,409; 7,246; 2,030; 8,925; 5,634.

   You can cross out 7,246 and 8,925.

   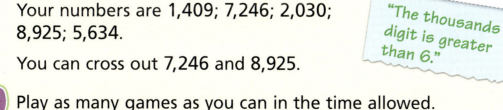

   thousands digit > 6

   "The thousands digit is greater than 6."

5. Play as many games as you can in the time allowed.

CHALLENGE

In 1858, units of money in the United States were different from those we use today. There were five units—mill, cent, dime, dollar, and eagle. The chart shows how much each unit was worth and its symbol.

United States Money in 1858
10 mills (m.) = 1 cent (c.)
10 cents (c.) = 1 dime (d.)
10 dimes (d.) = 1 dollar ($)
10 dollars ($) = 1 eagle (E)

Use the chart to answer the questions below. They come from a math textbook that was used by students in 1858!

1. How many mills in 2 cents?

2. How many cents in 3 dimes?

3. How many dimes in 4 dollars?

4. How many dollars in 2 eagles?

5. How many dimes in 1 eagle?

6. How many dimes in 3 dollars and 6 dimes?

7. How many cents in 4 dimes and 7 cents?

8. How many dimes are equal to 70 cents?

9. If James earned 12 dollars and his father earned 3 eagles, how many dollars did they earn together?

10. A man has 4 eagles, 4 dollars, and 4 dimes. How many dollars and cents does he have?

Chapter 5
Understanding Addition and Subtraction Algorithms

Dear Student,

An algorithm is a step-by-step process to solve a problem. You have already learned algorithms to add and subtract large numbers, and you have experience using them. This chapter will give you a closer look at how these algorithms work.

Look at these puzzles:

200	50	4	254
700	40	6	746
900	90	10	■

600	140	6	746
200	50	4	254
400	90	2	■

How are they like the problems shown below? How are they different from the problems below?

$$\begin{array}{r} 254 \\ + 746 \end{array} \qquad \begin{array}{r} 746 \\ - 254 \end{array}$$

By exploring questions like these, you will become even more skilled at computing, and will learn more about how arithmetic works.

Best wishes for a fun chapter!

Mathematically yours,
The authors of **Think Math!**

THE WORLD ALMANAC FOR KIDS

All Kinds of Puzzles

Puzzles challenge us to think and use our brains. Crossword puzzles teach us about words. Jigsaw puzzles help us tell different shapes apart. Did we forget to mention that puzzles are fun?

FACT·ACTIVITY 1

Melissa likes jigsaw puzzles, especially ones that are unusual shapes. All of Melissa's shaped puzzles have about 1,000 pieces. The chart shows the exact number of pieces in each of her shape puzzles.

Shape Puzzles	
Shape	Number of Pieces
Bear	899
Frog	975
Ladybug	918
Snowman	981
Space Shuttle	995
Statue of Liberty	958

Use the chart to answer the questions.

1. Which is the puzzle with the least number of pieces? Explain how you could represent the number using the fewest base-ten blocks.

2. If the puzzle with the most pieces takes you the longest to assemble, which puzzle would take you the most time to complete?

3. Which puzzle has more pieces, Snowman or Ladybug?

4. Write the number of puzzle pieces in order from least to greatest.

5. Melissa wants to organize her puzzles into those that are about 900 pieces and those that are about 1,000 pieces. Use rounding to tell which puzzles belong in each category.

FACT•ACTIVITY 2

Some puzzles come in books. Melissa has puzzle books and other types of books. She takes the following five books with her on a long trip.

For 1–4, use the table.

1. How much longer is the Crossword Puzzle Book than the Number Puzzle Book?

2. If Melissa reads the Funny Poems book and Princess Tale book by the end of her trip, how many pages will she have read?

3. Melissa has read 47 pages of the animal stories book. How many more pages must she read to finish the book?

4. What is a good estimate for the total number of pages in all of the books? Explain.

Melissa's Books

Title	Number of Pages
Number Puzzle Book	32
Crossword Puzzle Book	80
Animal Stories	96
Funny Poems	128
The Princess Tale	112

CHAPTER PROJECT

What kinds of books do you like? Some kinds of books are puzzle books, animal books, fantasy books, joke books, science books, and reference books.

- Use books from your home, school, or library. List 2 books in each of 3 different categories, such as 2 joke books, 2 science books, and 2 adventure books. You can use books that you have read or books that you would like to read.

- Make a poster with the name of each category, the titles of the books in each category, and the number of pages in each book.

- Find the total number of pages for the books in each category.

- Describe how you might find the total number of pages in all 6 books.

Some jigsaw puzzles have as many as 18,000 pieces.

Chapter 5
Lesson 1
REVIEW MODEL
Comparing Numbers

You can use base-ten blocks to help you compare numbers.

Example Compare 1,146 and 1,163.

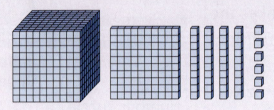 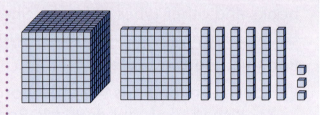

Step 1 Use base-ten blocks to show each number. Then compare from left to right.

Step 2 Look at the thousands. They are the same, so continue to compare.

Step 3 Look at the hundreds. They are the same, so continue to compare.

Step 4 Look at the tens. 6 tens is greater than 4 tens. So, 1,163 is the greater number. (You do not need to look at the ones, because the tens are different.)

Step 5 Use these symbols: < (less than) and > (greater than)

1,146 < 1,163   1,163 > 1,146

✓ Check for Understanding

Compare the numbers. Write < or >.

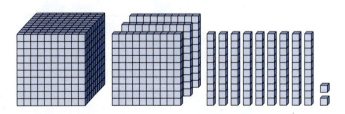   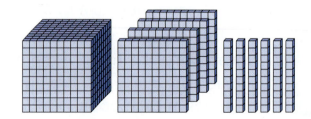

1,392 ● 1,460

70   Chapter 5

Chapter 5
Lesson 3
EXPLORE
Jumping on the Number Line

1 A frog jumped from **47 to 50** on the number line.

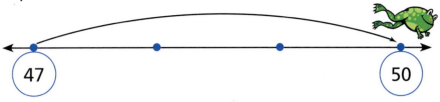

How many spaces did the frog jump?

2 Next, the frog jumped from **50 to 80**.

How far was the second jump?

3 Then the frog jumped from **80 to 82**.

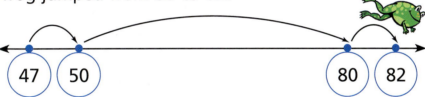

How far was the third jump?

4 How far is it to jump all the way from **47 to 82**?

5 Solve 82 − 47 = ■

6 Draw a picture showing one or more jumps to go from 66 to 104 on the number line. Find the distance for each jump you make and the total distance for all the jumps.

Chapter 5, Lesson 3
REVIEW MODEL
Using the Number Line to Find Differences

You can use the number line to help you subtract. The distance between the two numbers is their difference.

Example 47 − 18 = ■

Step 1 Draw the section of the number line that starts with the smaller number and ends with the larger number.

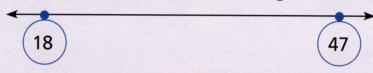

Step 2 Jump from the smaller number to the larger number. Use landing places that are easy to work with, such as multiples of ten.

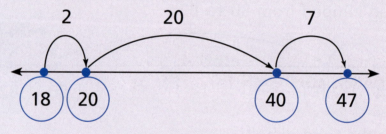

Step 3 Find the total distance jumped.

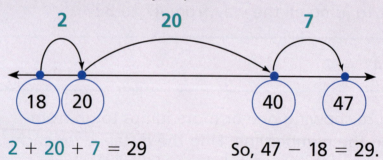

2 + 20 + 7 = 29   So, 47 − 18 = 29.

✓ Check for Understanding

Find the difference. Draw the number line on your own paper.

1 93 − 55 = ■   **2** 138 − 119 = ■

Chapter 5 Lesson 4
EXPLORE
Predicting Digits

Predict the number of base-ten rods you will need to show each sum with fewest blocks.

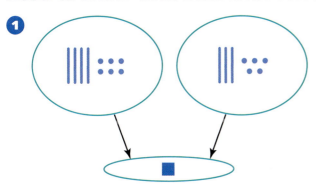
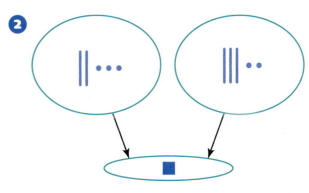

Write the answer to each question.

3. Will **28 + 45** be in the **sixties** or the **seventies**?

4. Will **16 + 78** be in the **eighties** or the **nineties**?

Predict the number of base-ten flats you will need to show each sum with fewest blocks.

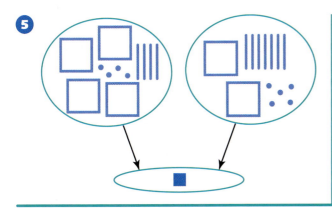
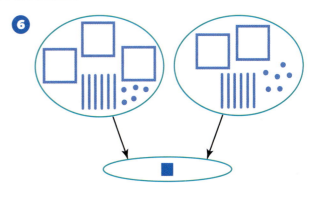

Write the answer to each question.

7. Will **356 + 482** be in the **700s** or the **800s**?

8. Will **238 + 319** be in the **500s** or the **600s**?

Chapter 5
Lesson 5
EXPLORE
Predicting Sums

Imagine combining the two piles of base-ten blocks and making trades until you have the fewest blocks for the sum.

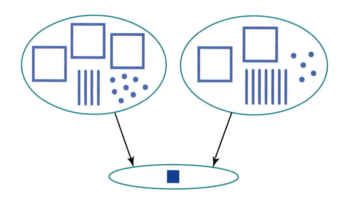

1 How many flats will be in the sum?

2 How many rods will be in the sum?

3 How many units will be in the sum?

4 Add to check your predictions.
$$\begin{array}{r} 348 \\ + 275 \\ \hline \end{array}$$

Without calculating $\begin{array}{r} 183 \\ + 594 \\ \hline \end{array}$ **, predict each digit of the sum.**

5 the hundreds digit

6 the tens digit

7 the ones digit

8 Add to check your predictions.
$$\begin{array}{r} 183 \\ + 594 \\ \hline \end{array}$$

Chapter 5
Lesson 5

REVIEW MODEL
Addition with Regrouping

You can use base-ten block pictures to find a sum. Regroup to find the fewest blocks.

Step 1
Look at the hundreds.

$$\begin{array}{r} 137 \\ + 385 \end{array}$$

Step 2
See if you have 10 tens you can regroup to make another hundred.

Step 3
See if you have 10 ones you can regroup to make another ten.

Step 4
Count the remaining blocks. There are 5 hundreds, 2 tens, and 2 ones. So,

$$\begin{array}{r} 137 \\ + 385 \\ \hline 522 \end{array}$$

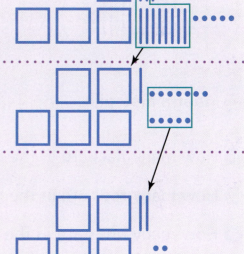

✓ Check for Understanding

Find the sum. Draw the base-ten blocks on your own paper.

1. $\begin{array}{r} 214 \\ + 269 \end{array}$

2. $\begin{array}{r} 452 \\ + 178 \end{array}$

Chapter 5
Lesson 8
EXPLORE
Predicting Differences

Tuan needed to split his collection of base-ten blocks into two piles. In order to put the blocks he wanted into the first pile, he had to trade some of his original blocks for smaller ones. Show what is left for the other pile.

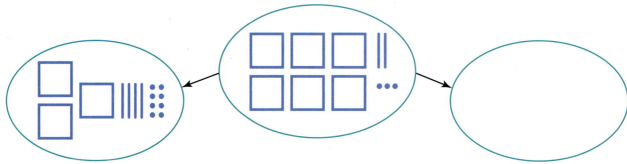

Tuan's first pile   Tuan's original collection   Tuan's second pile (the difference)

1 How many flats are in the difference?

2 How many rods are in the difference?

3 How many units are in the difference?

4 Subtract to check your predictions.

$$\begin{array}{r} 623 \\ -348 \\ \hline \end{array}$$

Without calculating $\boxed{\begin{array}{r} 777 \\ -594 \\ \hline \end{array}}$, predict each digit of the difference.

5 the hundreds digit

6 the tens digit

7 the ones digit

8 Subtract to check your predictions.

$$\begin{array}{r} 777 \\ -594 \\ \hline \end{array}$$

Chapter 5, Lesson 9

REVIEW MODEL

What Is a Cross Number Puzzle?

Cross Number Puzzles are a tool for adding and subtracting multi-digit numbers.

Three-digit numbers are separated into hundreds, tens, and ones, so you can add or subtract each place value.

Addition Puzzle

345
+ 564

Hundreds	Tens	Ones	
300	40	5	345
500	60	4	564
800	100	9	909

Subtraction Puzzle

891
− 704

Hundreds	Tens	Ones	
800	80	11	891
700	0	4	704
100	80	7	187

> Use 80 and 11, so 4 can be combined with something (7) to get 11, rather than using 90 and 1 and trying to combine 4 with something to get 1.

Amounts on either side of a heavy line must be the same.

Addition Puzzle

| 300 | 40 | 5 | 345 | 300 + 40 + 5 = 345
| 500 | 60 | 4 | 564 | 500 + 60 + 4 = 564
| 800 | 100 | 9 | 909 | 800 + 100 + 9 = 909

```
  300     40     5     345
+ 500   + 60   + 4   + 564
  800    100    9     909
```

Subtraction Puzzle

| 800 | 80 | 11 | 891 | 800 + 80 + 11 = 891
| 700 | 0 | 4 | 704 | 700 + 0 + 4 = 704
| 100 | 80 | 7 | 187 | 100 + 80 + 7 = 187

```
  700      0     4     704
+ 100   + 80   + 7   + 187
  800     80    11    891
```

✓ Check for Understanding

Copy the Cross Number Puzzle on your own paper. Then complete the puzzle.

1.

400	20	3	■
100	50	9	■
■	■	■	■

2.

500	140	■	647
■	50	6	256
■	■	■	■

Chapter 5   77

Chapter 5, Lesson 11

REVIEW MODEL
Problem Solving Strategy
Solve a Simpler Problem

> The volunteers made 430 care packages to send overseas. They mailed 249 packages on Monday. Do they have enough packages left to meet their goal of mailing at least 200 packages on Tuesday?

Strategy: Solve a Simpler Problem

Read to Understand

What do you know from reading the problem?

There were 430 care packages made and 249 packages were mailed on Monday. You need to find out if there are 200 packages left.

Plan

How can you solve this problem?

You can solve a simpler problem.

Solve

How can you solve a simpler problem?

You do not need to subtract the ones, tens, and hundreds to find the exact difference. You just need to find how many hundreds are left.

$$\begin{array}{r} 430 \\ -\ 249 \\ \hline \end{array}$$

Write a subtraction problem and work from left to right to find the number of hundreds in the difference. Look at the tens digits to see if you need to regroup a hundred to subtract the tens. You cannot subtract 4 tens from 3 tens, so you will need to trade 1 hundred for 10 tens. If you subtract 2 hundreds from the 3 hundreds left, you see they will not meet their goal of mailing 200 packages on Tuesday.

Check

Look back at the problem. Did you answer the question that was asked? Does the answer make sense?

Problem Solving Practice

Use the strategy *solve a simpler problem*.

1) It takes Yolanda 1 minute to copy 10 pages. She made 46 copies of one page and 52 copies of another page. Did Yolanda finish making the copies in 10 minutes?

2) Jeremy has $2.00. He wants to buy a cup of yogurt for 89¢ and 2 pieces of fruit for 48¢ each. Does Jeremy have enough money?

Problem Solving Strategies
- ✔ Act It Out
- ✔ Draw a Picture
- ✔ Guess and Check
- ✔ Look for a Pattern
- ✔ Make a Graph
- ✔ Make a Model
- ✔ Make an Organized List
- ✔ Make a Table
- ✔ **Solve a Simpler Problem**
- ✔ Use Logical Reasoning
- ✔ Work Backward
- ✔ Write a Number Sentence

Mixed Strategy Practice

Use any strategy to solve. Explain.

3) An artist can make 6 clay pots in 2 days. How many clay pots can he make in 8 days?

4) There are 36 birds in a special exhibit at the zoo. There are 8 more females than males. How many birds are male?

5) A road map shows the Baker family must travel 267 miles south and then 156 miles west to get to the campground. How far must they travel?

6) Josh brought home 3 watermelons left at the end of the picnic. There were 14 watermelons eaten at the picnic. Tory took 2 watermelons home. How many watermelons were there at the beginning of the picnic?

7) Marissa is making a picture frame. She glues stones across the top of the frame. She continues the pattern along the bottom. Look at the picture. What color stone will Marissa place next?

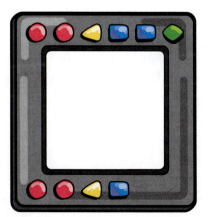

Chapter 5 Vocabulary

Choose the best vocabulary term from Word List A for each sentence.

Word List A
- difference
- estimate
- exchange
- group
- multiple
- number line
- predict
- round
- sum

1. A(n) __?__ is a line labeled with evenly spaced numbers.

2. To find a(n) __?__ of a sum, add rounded numbers instead of the exact numbers.

3. To find the __?__ between two numbers, subtract one number from the other.

4. The number 15 is a __?__ of 3.

5. Trade means the same as __?__.

6. To find the __?__ of two numbers, add the numbers.

7. To __?__ a number to the nearest hundred, decide which multiple of 100 the number is closer to.

Complete each analogy using the best term from Word List B.

Word List B
- unit
- regroup
- rod
- sum

8. Flat is to 100 as __?__ is to 10.

9. Addition is to __?__ as subtraction is to difference.

Talk Math

Discuss with a partner what you have learned about addition and subtraction. Use the vocabulary terms *group* and *regroup*.

10. How can you use flats, rods, and units to find the total for 2 three-digit numbers?

11. How do you know when you need to trade to find a difference?

Word Line

12 Create a word line for the words *unit*, *flat*, and *rod*.

Words:

Sequence:
_____

Venn Diagram

13 Create a Venn diagram for addition and subtraction. Include these terms: *base-ten blocks, Cross Number Puzzle, difference, estimate, number line, regroup, sum,* and *total.*

What's in a Word?

TRADE The word *trade* has many different meanings. *Trade* can describe a job. "He is a car mechanic by *trade*." It can also be used to mean "business." "She did an excellent *trade* in souvenirs for the fair." It can be a naming word for an exchange. "He made a good *trade*." It can be an action word for an exchange. "She and her friend *traded* jackets." In math, you can *trade* hundreds for tens or tens for ones when subtracting. If you use base-ten blocks, you can *trade* or exchange 1 rod for 10 units.

GO ONLINE Technology
Multimedia Math Glossary
www.harcourtschool.com/thinkmath

Chapter 5

GAME

Ordering Numbers

Game Purpose
To practice comparing and ordering four-digit numbers

Materials
- 4 number cubes (labeled 1–6)

1,246	1 point
1,246	1 point
4,612	2 points
6,412	3 points

How to Play the Game

1 This is a game for 4 to 6 players. The goal is to score points by making up numbers from a toss of the number cubes.

2 The group tosses all 4 number cubes. Each player uses the numbers tossed to secretly write a four-digit number.

3 Everyone shows their numbers. Work together to put the numbers in order from smallest to largest.

4 This is how you earn points:
- 2 points if no one else has the number
- 1 point for the smallest number
- 1 point for the largest number

Example: Number cubes

Player	Number	Points Earned	Total Points
Carli	5,265	No one else has it: 2 points Smallest number: 1 point	3
Lamar	5,562	No one else has it: 2 points	2
Royce	6,552	Largest number: 1 point	1
Becka	6,552	Largest number: 1 point	1

5 Keep playing until time is called. The player with the most points wins. Ties are possible.

Least to Greatest

Game Purpose
To practice estimating and ordering sums

Materials
- Activity Masters 33 and 49: Least to Greatest Cards
- Stopwatch or a watch with a second hand

How to Play the Game

1. Play this game with a partner. Cut out the cards from Activity Masters 33 and 49. Mix them up. Put them face down in a pile.

2. Choose one player to be the card Placer. The other player will be the Timer.

3. The Timer says, "Go." The Placer turns over one card at a time and puts it where it belongs in a line from the least sum to the greatest sum. You can pass on a card by setting it aside.

4. The Timer says "Stop" after 60 seconds and checks the order of the cards. The Timer says whether there are mistakes but not what the mistakes are.

5. The Placer may remove cards from the line to correct the order. When the Timer agrees that the order is correct, the Placer gets 1 point for each card.

   Example: These four cards are placed correctly.

   | 18 | 14 | 481 | 360 |
   |+36 |+68 |+ 23 |+386|

6. Switch roles. Play until time is called. The player with more points wins!

CHALLENGE

Here is a math trick that will let you add 3 three-digit numbers in your head very quickly.

Practice this trick on your own. Then try it on friends and family members.

Step 1 Ask someone to name a three-digit number. Suppose the person says 534. This number is the key to the answer. Write the number. — 534

Step 2 Ask for a second three-digit number. Suppose the person says 741. Write it below the first one. — 741

Step 3 Then write the third number. You want the second and third numbers to have a sum of 999.
Think: 7 + 2 = 9, 4 + 5 = 9, and 1 + 8 = 9.
So, you write 258. — 258
The addition looks like this:

```
  5 3 4    First number
  7 4 1    Second number
+ 2 5 8    Third number
```

Step 4 Now you can write the sum without adding the columns. Here's how.
Think: 741 + 258 = 999, which is **1 less than 1,000.**
Think: 1,000 + 534, or 1,534. That is why the first number is the key to the answer. The sum is **1,000 + the first number − 1.** And you can do that using mental math! — 1,533

Practice the trick. Find the third number and the sum for each set of numbers without adding the columns. Then use a calculator to check.

1 428, 375   **2** 856, 602   **3** 787, 529

84 Chapter 5

Chapter 6
Rules and Patterns

Dear Student,

The number of clouds on the back of this card was found by applying a rule to the number of clouds on the front. What do you think the rule could be?

Some ideas for the rule might be:

- Multiply the number of clouds on the front by 2.
- Add 1 cloud to the amount on the front.
- Add 3 clouds, and then subtract 2 clouds from the clouds on the front.

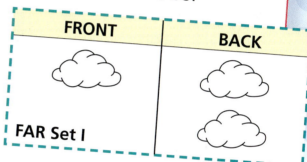

FAR Set I

You cannot be sure of the rule until you see the other cards in the set that follow the same rule. In this chapter, you will be working with sets of cards like these as you think about patterns. Have fun finding a rule for all the cards!

Mathematically yours,
The authors of *Think Math!*

THE WORLD ALMANAC FOR KIDS

Beading Fun

For thousands of years, beads have been used for jewelry, decorations, and trade. Beads can be made from shells, gems, glass, and even from seeds.

FACT·ACTIVITY 1

Two girls are making necklaces. Alona uses 3 white beads for every red bead she puts in her necklace. Cara uses 3 more small beads than the number of large beads in her necklace.

| Alona's Necklace ||
Red	White
1	3
2	6
3	9
4	■
5	■

| Cara's Necklace ||
Large	Small
2	5
4	7
6	9
8	■
10	■

Use the tables to answer the questions.

1. What is a rule for finding the number of white beads in Alona's necklace for a particular number of red beads? How many white beads are needed if there are 4 red beads? 5 red beads?

2. Look at the table for Cara's Necklace. What is a rule for finding the number of small beads in Cara's necklace for a particular number of large beads? What are the missing numbers in the table?

86 Chapter 6

FACT·ACTIVITY 2

Janet is stringing beads to make a necklace. The table below shows the relationship between the number of long purple beads and round yellow beads.

🟪	2	4	■	8	■	■	■
🟡	3	6	9	■	■	■	■

Use the table to answer the questions.

1. Janet begins with 2 purple beads and 3 yellow beads. When there are 4 purple beads, how many yellow beads are there?
2. When there are 9 yellow beads, how many purple beads are there?
3. Complete the table.
4. Draw a picture of what Janet's necklace might look like. Describe the pattern.

CHAPTER PROJECT

Look in magazines or catalogs for floor tiles that have simple geometric shapes. You may find something like the design below.

- Select a tile or draw a design that has a clear number of two or more geometric shapes on it.
- Suppose you have 2, 3, 4, or 5 of the tiles. Write a rule that gives the number of each shape when you know how many tiles you have.
- Make a table to show the relationship between the shapes.

square

trapezoid

ALMANAC Fact

The Bead Museum in Washington, D.C., has exhibits showing how beads have been used throughout history and amazing beaded crafts from all over the world. The largest bead in the collection is 6 inches long and weighs 6 pounds.

Chapter 6 Lesson 2
REVIEW MODEL
Recording Data

You can record information from Find a Rule cards onto a graph.

The numbers on the front of the FAR cards refer to the number of stickers. The numbers on the back of the cards refer to the cost of the stickers.

FRONT	BACK
1	10¢

FRONT	BACK
2	20¢

Step 1

Look at the first card to find the first number of stickers. On the graph, find the line for 1 sticker.

Step 2

Look at the first card to find the cost of the sticker. On the graph, find the line labeled 10¢.

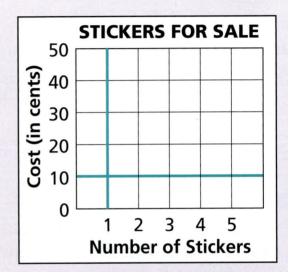

Step 3

Place a point where the two lines intersect.

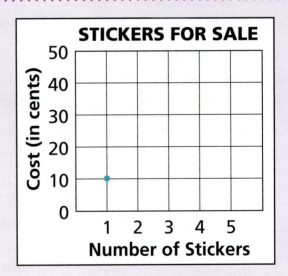

✓ Check for Understanding

1. Use the second FAR card at the top of the page. Explain how you would record the information on the graph above.

88 Chapter 6

Chapter 6
Lesson 4

EXPLORE
Exploring FAR Cards

These FAR cards have two rules.

A

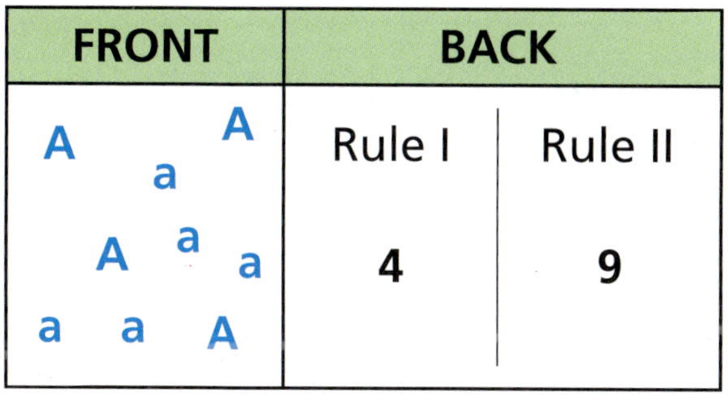

FRONT	BACK	
A A a A a a a a A	Rule I 4	Rule II 9

B

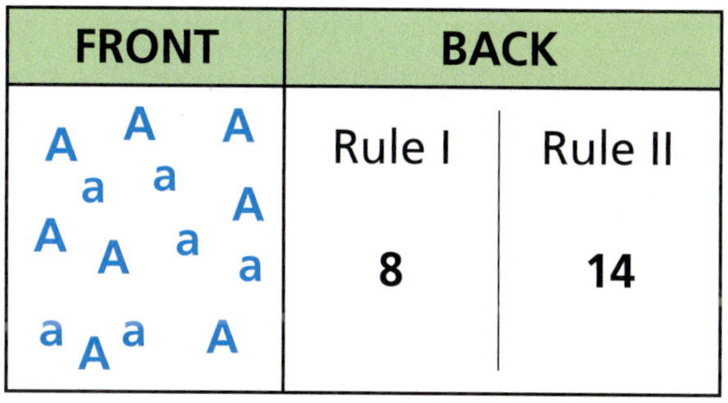

FRONT	BACK	
A A A a a A A a A a a A	Rule I 8	Rule II 14

C

FRONT	BACK	
A A A A a A A A A	Rule I 8	Rule II 9

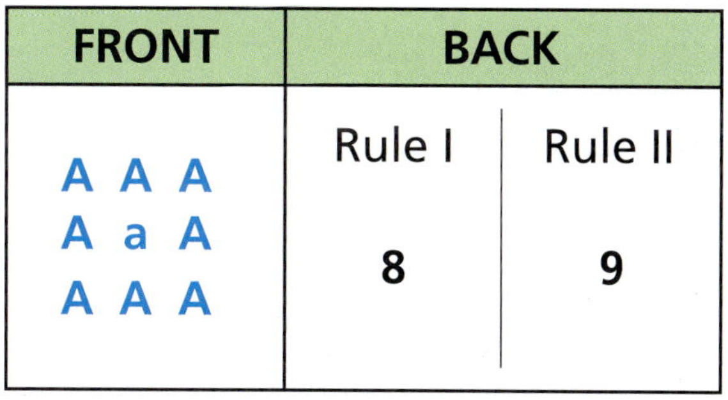

This table shows the numbers on the backs of cards A, B, and C.

Card	Rule I	Rule II
A	4	9
B	8	14
C	8	9

On your own paper, continue the table for cards D, E, and F. The fronts of those cards are shown below.

D

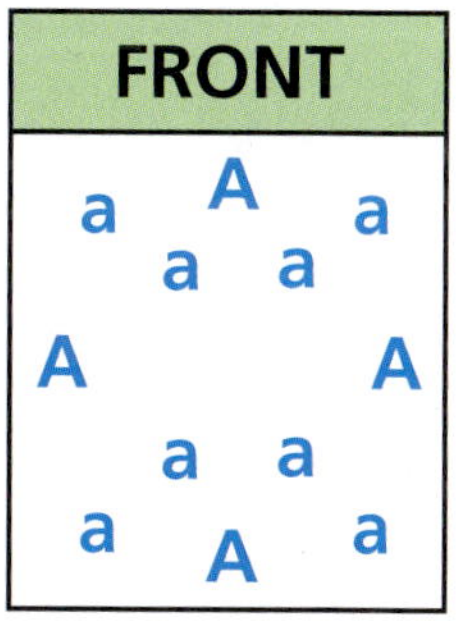

FRONT
a A a a a A A a a a A a

E

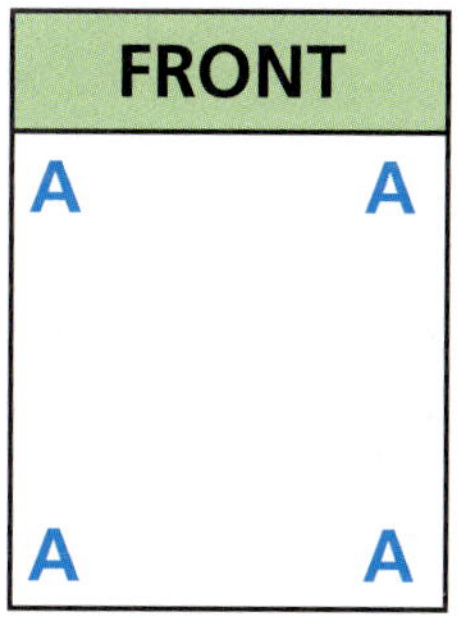

FRONT
A　　　A   A　　　A

F

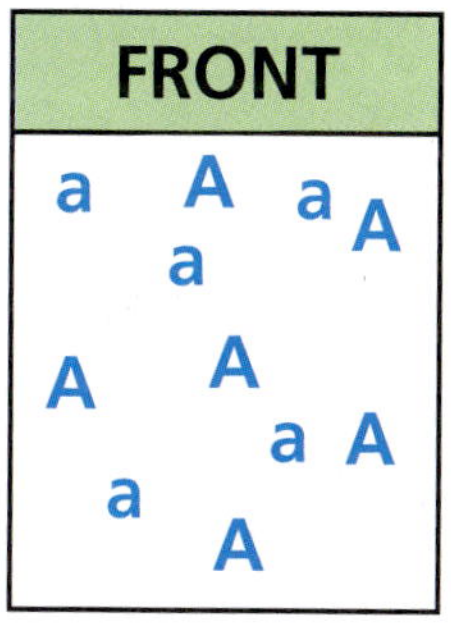

FRONT
a A a A a A A a A a A

Chapter 6　**89**

Chapter 6 Lesson 6
EXPLORE
Exploring a Pattern

Study the figures below. Look for a pattern.

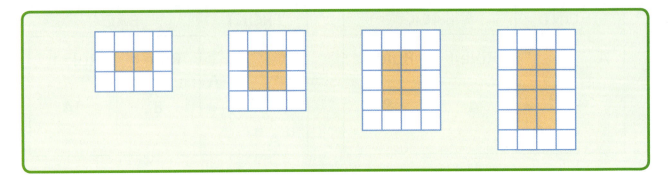

1 How many tiles are in each figure?

2 How many orange tiles are in each figure?

3 How many white tiles are in each figure?

4 How does the total number of tiles change from one figure to the next figure?

5 How does the number of orange tiles change from one figure to the next figure?

6 How does the width of the figure change?

7 How does the length of the figure change?

8 Draw the next figure.

90 Chapter 6

Chapter 6 Lesson 6

REVIEW MODEL
Finding a Rule

You can use a rule to describe the pattern in a sequence.

Step 1 Think about how each figure in the sequence is the same.

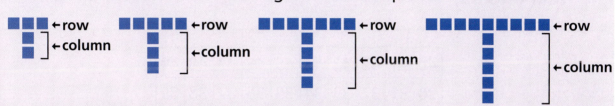

Each figure has a row with an odd number of tiles across the top and a column of tiles down the middle.

Step 2 Think about how each figure is different from the previous figure.

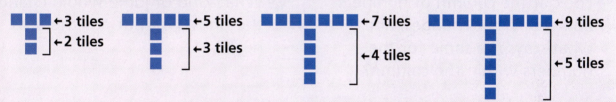

Each figure has two more tiles in the row across the top. Each figure has one more tile in the middle column.

Step 3 State the rule and draw the next figure.

Add two more tiles to the top row. Add one more tile to the middle column.

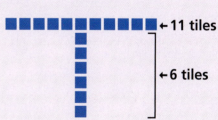

✓ Check for Understanding

Draw the next figure following the pattern.

1

2

Chapter 6   91

Chapter 6
Lesson 7

EXPLORE
Exploring the Number Line Hotel

A part of the Number Line Hotel is shown below. The entire hotel contains the number line from 0 to 99.

40	41	42	43	44	45	46	47	48	49
30	31	32	33	34	35	36	37	38	39
20	21	22	23	24	25	26	27	28	29
10	11	12	13	14	15	16	17	18	19
0	1	2	3	4	5	6	7	8	9

1 Look at the column of numbers above 6. What changes, and what stays the same for the numbers within the column?

2 What kind of jump would change the ones digit but not the tens digit?

3 How could you use the Number Line Hotel to help find 11 + 37?

4 How could you use the Number Line Hotel to help find 25 + 16?

5 How could you use the Number Line Hotel to help find 28 − 13?

6 How could you use the Number Line Hotel to help find 32 − 26?

Chapter 6 Lesson 7
REVIEW MODEL
Adding and Subtracting on a Grid

You can add and subtract by using a grid.

40	41	42	43	44	45	46	47	48	49
30	31	32	33	34	35	36	37	38	39
20	21	22	23	24	25	26	27	28	29
10	11	12	13	14	15	16	17	18	19
0	1	2	3	4	5	6	7	8	9

You can use an arrow to represent a move of 1 square on the grid.

↑: up   ↓: down   →: right   ←: left

To Add

Move up to add tens.
Move to the right to add ones.

1 move up (↑) adds 10.
So, 26 + 10 = 36.
3 moves right (→→→) adds 3.
So, 14 + 3 = 17.

To Subtract

Move down to subtract tens.
Move to the left to subtract ones.

4 moves down (↓↓↓↓) subtracts 40.
So, 45 − 40 = 5.
1 move left (←) subtracts 1.
So, 14 − 1 = 13.

Add tens and ones by using a combination of moves.

Start on 13. Move up 3 (↑↑↑).
Move 2 to the right (→→).
You land on 45.
So, 13 + 32 = 45.

Subtract tens and ones by using a combination of moves.

Start on 37. Move down 2 (↓↓).
Move 3 to the left (←←←).
You land on 14.
So, 37 − 23 = 14.

✓ Check for Understanding

Write the landing number. Then write an addition or subtraction sentence to match.

1 15↑↑→ ■   **2** 7↑↑↑→→ ■   **3** 43↓←←← ■

Chapter 6
Lesson 8
EXPLORE
Exploring Sharing Machine A

This machine takes a package as input. It outputs two smaller packages that share the contents of the input equally between them. Together, these two smaller packages contain everything the input package contained.

Aki put a package containing 2 quarters, 4 dimes, and 2 pennies into the machine. Use coins to act out what the machine will do.

1 What was in each package that came out of the machine?

2 How much money did the input package contain?

3 How much money did each output package contain?

4 Aki then tried an input package of 6 dimes, 2 nickels, and 4 quarters. What came out of the machine?

5 What did Aki put in the machine if the output was two packages, each containing 2 quarters and 4 pennies?

6 Aki input a package of 18 marbles. What came out of the machine?

7 The machine returns all packages that it cannot share evenly without cutting an object. What might you put in the machine that it would return?

Chapter 6 Lesson 8
REVIEW MODEL
Writing Division Sentences

You can write division number sentences to represent sharing situations.

There are 8 markers.

There are 15 trading cards.

The markers are shared equally between 2 students.

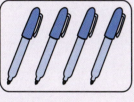 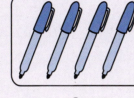

     1   2

Each student gets 4 markers.
So, 8 ÷ 2 = 4.

The cards are shared equally among 3 friends.

Each person gets 5 cards.
So, 15 ÷ 3 = 5.

✓ Check for Understanding

Write a division sentence for each situation.

1

There are 9 star stickers. They are shared equally among 3 third graders.

2

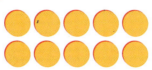

There are 10 counters. They are shared equally between 2 groups.

3

There are 12 quarters. They are shared equally among 3 brothers.

Chapter 6
Lesson 11

REVIEW MODEL
Problem Solving Strategy
Look for a Pattern

Leo makes a table to show how many stickers each of his brothers will get if he shares different numbers of stickers equally among them. How many stickers will each brother get if Leo shares 36 stickers?

Number of Stickers	12	4	8	28	20
Number Each Brother Gets	3	1	2	7	5

Strategy: Look for a Pattern

Read to Understand

What do you need to find out?

You need to find out how many stickers each brother will get if Leo shares 36 stickers.

Plan

How can you solve the problem?

You can look for a pattern.

Solve

How can you look for a pattern to solve the problem?

Look for a pattern in the table. The rule is divide by 4; 36 ÷ 4 = 9; So, each brother will get 9 stickers.

Check

Look back at the problem. Did you answer the question that was asked? Does the answer make sense?

Problem Solving Practice

Look for a pattern to solve.

1. Robert uses a pattern to stack boxes for a store display. He puts 11 boxes in the first row. He puts 9 boxes in the second row and 7 boxes in the third row. How many rows of boxes will be in the display when Robert is finished, if the top row has only 1 box?

2. Noriko uses a pattern to write the following numbers: 5 10 20 40
What is a rule for Noriko's pattern?

Problem Solving Strategies

- ✓ Act It Out
- ✓ Draw a Picture
- ✓ Guess and Check
- ✓ **Look for a Pattern**
- ✓ Make a Graph
- ✓ Make a Model
- ✓ Make an Organized List
- ✓ Make a Table
- ✓ Solve a Simpler Problem
- ✓ Use Logical Reasoning
- ✓ Work Backward
- ✓ Write a Number Sentence

Mixed Strategy Practice

Use any strategy to solve. Explain.

3. Jenna is choosing an outfit to wear. She can wear a long-sleeve shirt or a short-sleeve shirt. She can wear a skirt, short pants, or long pants. How many different outfits are possible?

4. Tyrone earned some money doing chores. He put half of what he earned in the bank. Then he paid his sister the $2 he owed her. He has $5 left to spend. How much did Tyrone earn doing chores?

5. There are 6 people in the Ling family. At the mall, each person in the family bought either a pizza for $4 or a hot dog for $3. They spent $22 in all. How many pizzas and hot dogs did the Ling family buy?

6. Eduardo makes snack bags for a class trip. He has 34 pretzels. If he puts 3 pretzels in each bag, how many bags can Eduardo fill?

7. Jacob and Gabe plan to go fishing every week this summer. The table shows how many fish the boys have caught each week so far. If the pattern continues, how many fish will they catch in Week 5?

FISH CAUGHT					
Week	1	2	3	4	5
Number of Fish	9	12	15	18	

Chapter 6 Vocabulary

Choose the best vocabulary term from Word List A for each sentence.

1. 6 ÷ 3 = 2 is a __?__.

2. A(n) __?__ can be made from two numbers and an operation symbol.

3. The number that is being divided is called the __?__.

4. A(n) __?__ is a picture of information.

5. In 45 ÷ 5 = 9, the number 5 is the __?__.

6. The number on the bottom of a fraction is called the __?__.

7. A table can be used to show the __?__ for an input.

Word List A

column
denominator
dividend
division sentence
divisor
expression
graph
input
numerator
output
pattern
row
rule

Complete each analogy. Use the best term from Word List B.

8. Numerator is to denominator as __?__ is to whole.

9. Addition is to sum as division is to __?__.

Word List B

dividend
package
part
quotient

💬 Talk Math

Discuss with a partner what you know about rules and patterns. Use the vocabulary terms *rule* and *pattern*.

10. You want to write a rule for a table. How do you know which operation to use?

11. How can you find a rule from a graph?

Word Definition Map

12 Create a word definition map for the word *graph*.

  A What is it?

  B What is it like?

  C What are some examples?

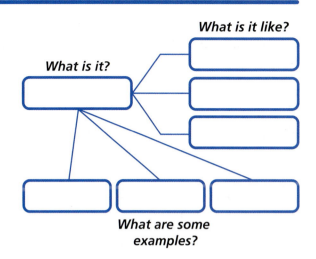

Tree Diagram

13 Create a tree diagram using the word *fraction*. Use what you know about the words *numerator, denominator, top number, bottom number, part,* and *whole.*

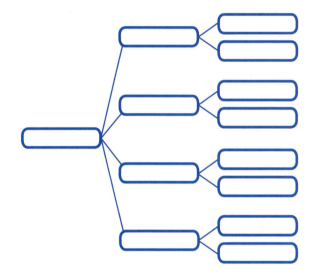

TABLE A *table* is a piece of furniture with a smooth flat top fixed on legs. Some people use a table as a desk or a workbench. A *table of contents* in a book is a short list of what is in the book and where to find it. In math, a table uses columns and rows to organize and display information like a table of contents. The information can be numbers, but it does not have to be.

GO ONLINE Technology
Multimedia Math Glossary
www.harcourtschool.com/thinkmath

GAME

Find a Rule

Game Purpose
To practice using two-number inputs and outputs to find a rule

Materials
- Activity Master 68: Find a Rule

How to Play the Game

1. Play this game with a partner. Each player thinks of a rule for two input numbers and keeps it a secret.

2. Your partner names pairs of inputs. Write them in the first column of your table. Then use mental math or paper and pencil along with your rule to find the outputs.

3. When you have filled in the rows of the table, your partner guesses the rule and scores 1 point if correct.

4. Switch roles and continue playing. The winner is the player with more points when time is called.

Find a Rule

Two Numbers	Rule A	Rule B
2, 3	6	
10, 9	90	
4, 2	8	
7, 4	28	
3, 11	33	
8, 8	64	

GAME

Make a Rule

Game Purpose
To practice finding a rule for a set of numbers

Materials
- Index cards

How to Play the Game

1. Play this game with a partner. Make 2 sets of number cards, each numbered 1 through 12. Mix up all the cards. Place them face down in a stack.

2. Take turns turning over a card from the top of the deck. Both players look for 3 cards on the table that follow a rule.
   - Rules such as "greater than" and "less than" are not allowed.
   - A rule can be used only once. For example, if the rule for a set of cards is "numbers that are even," that rule cannot be used for another set of cards.

3. The first player to correctly name a rule for 3 cards takes the cards. Keep a record of the rules you use.

4. If neither player can name a rule, the cards remain on the table.

   Example: The number cards are: 8, 8, 12

   Possible rules are:
   - even numbers
   - multiples of 4
   - one more than an odd number
   - one less than an odd number

5. Play until all the cards in the stack have been turned over and neither player can find a rule for the cards that are left. The winner is the player with more cards.

Chapter 6

CHALLENGE

Put your finger on one corner of Picture A. Can you trace a path over the picture without going back over any line? Try it. A path that works is shown on the right.

Picture A

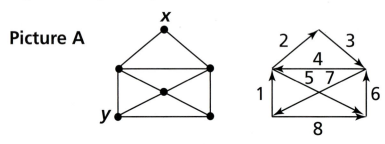

Pictures like this one are called *networks*. They have even and odd corners. Corner *x* is even, because it has an even number of lines meeting at the dot. Corner *y* is odd, because it has an odd number of lines meeting at the dot.

Trace over each picture. Try not to go back over a line.

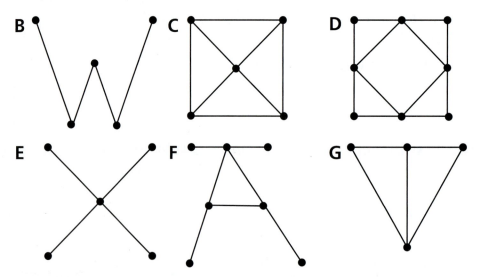

Use the table to help you look for a pattern.

Picture	A	B	C	D	E	F	G
How many odd corners?	2	■	■	■	■	■	■
Can you trace over it?	yes	■	■	■	■	■	■

You can trace over a picture if it has ■ or ■ odd corners.

Chapter 7 Fractions

Dear Student,

You have seen fractions in many places: when you need one third of a cup of flour, when it's half an hour until lunch, and when you buy something with a quarter (one fourth of a dollar). In this chapter, you will not only look at fractions and what they mean, but you will also compare fractions and develop strategies for thinking about them.

An important idea about fractions is they are made by cutting a whole into equal parts. How could you cut this rectangle into halves (2 pieces with the same area)? How could you cut this rectangle into 3 pieces with the same area?

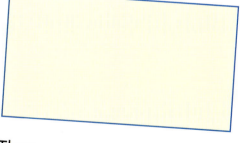

There are many ways to cut a rectangle into halves and thirds, so be creative!

Mathematically yours,
The authors of **Think Math!**

Paper Folding Fun

FACT·ACTIVITY 1

Origami is a Japanese word that means "paper folding." There are 6 basic folds in origami: the mountain fold, valley fold, diagonal fold, fold and unfold, rotate, and flip over. These basic folds are used to create complex origami designs.

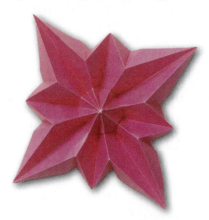

The models on the right show some folds that create fractional parts. Use these models for Problems 1–4.

1. What fractional parts are shown in Model A?
2. What fraction describes 3 of the equal parts in Model B?
3. How can you fold Model C to show fourths? Explore by folding a piece of square paper. Unfold your paper to show your folds. Draw a picture to show your answer.
4. Fold Model C into eighths. Show a fraction that is equivalent to your answer in Problem 2 by coloring some of the folded parts.

A

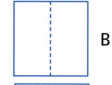

B

C

FACT·ACTIVITY 2

Imagine you are making origami designs to sell at the school arts and crafts fair. The table shows the time it takes to make each design and how much each design will cost.

Origami Designs		
Item	Time (minutes)	Price
bookmark	15	25¢
crane	10	30¢
coaster	20	50¢
frog	30	75¢

Use a clock or coins for help.

1. Write each time in the table as a fraction of an hour.

2. Write each price in the table as a fraction of a dollar.

CHAPTER PROJECT

You can make more colorful origami by decorating the paper. Start with 2 pieces of white square paper. Fold each paper into 8 equal parts.

- On the first paper, create an interesting origami square by coloring one of the sections. Repeat the design on a number of sections so that your colored sections cover less than $\frac{5}{8}$ but more than $\frac{1}{4}$ of the paper.

- On the second paper, use three different colors on different sections of the square. Only use one color in each section. Write a fraction to show what part of the square each color represents.

- Use the papers you designed and fold each one into an object.

ALMANAC Fact

The world's largest origami crane was created in Japan to promote world peace. The crane was 120 feet tall. About 10,000 children from all over the world helped to create drawings on the paper.

Chapter 7 Lesson 1
REVIEW MODEL
Understanding Fractions

A fraction names a part of a whole.

The number below the line tells how many equal-size pieces the whole was cut into. The number above the line tells how many of those pieces you are referring to.

Example What fraction of the rectangle is shaded?

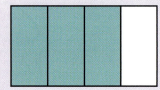

3 ← number of pieces shaded
―
4 ← number of equal-size pieces the whole was cut into

The **size** of each piece must be the same.
The **shape** of each piece does NOT have to be the same.

Examples

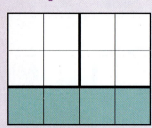 $\frac{1}{3}$ is shaded.

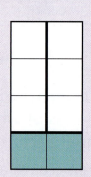 The shaded part is NOT $\frac{1}{3}$.

✓ Check for Understanding

Write *yes* or *no* to answer the question.

Is $\frac{1}{2}$ shaded?

 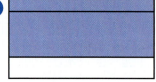
Is $\frac{2}{3}$ shaded?

 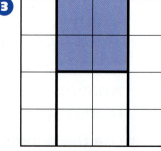
Is $\frac{1}{4}$ shaded?

Chapter 7
Lesson 2
REVIEW MODEL
Finding Equivalent Fractions

Equivalent fractions are two or more fractions that name the same amount.

You can use fraction models to help you find equivalent fractions.

Example Find fractions equivalent to $\frac{1}{4}$.

Find the models that match the length of $\frac{1}{4}$.

1											
$\frac{1}{2}$						$\frac{1}{2}$					
$\frac{1}{3}$				$\frac{1}{3}$				$\frac{1}{3}$			
$\frac{1}{4}$			$\frac{1}{4}$			$\frac{1}{4}$			$\frac{1}{4}$		
$\frac{1}{5}$		$\frac{1}{5}$		$\frac{1}{5}$			$\frac{1}{5}$		$\frac{1}{5}$		
$\frac{1}{6}$		$\frac{1}{6}$		$\frac{1}{6}$		$\frac{1}{6}$		$\frac{1}{6}$		$\frac{1}{6}$	
$\frac{1}{8}$		$\frac{1}{8}$		$\frac{1}{8}$	$\frac{1}{8}$		$\frac{1}{8}$	$\frac{1}{8}$		$\frac{1}{8}$	$\frac{1}{8}$
$\frac{1}{10}$	$\frac{1}{10}$	$\frac{1}{10}$	$\frac{1}{10}$	$\frac{1}{10}$	$\frac{1}{10}$	$\frac{1}{10}$	$\frac{1}{10}$	$\frac{1}{10}$	$\frac{1}{10}$		
$\frac{1}{12}$	$\frac{1}{12}$	$\frac{1}{12}$	$\frac{1}{12}$	$\frac{1}{12}$	$\frac{1}{12}$	$\frac{1}{12}$	$\frac{1}{12}$	$\frac{1}{12}$	$\frac{1}{12}$	$\frac{1}{12}$	$\frac{1}{12}$

$\frac{2}{8}$ and $\frac{3}{12}$ are the same length as $\frac{1}{4}$, so they are equivalent.

✓ Check for Understanding

Use the models above to write an equivalent fraction.

1. $\frac{1}{3}$

2. $\frac{1}{5}$

3. $\frac{1}{2}$

4. $\frac{5}{6}$

Chapter 7
Lesson 4 — EXPLORE: Cracked Eggs

What are some different ways to describe a half dozen?

 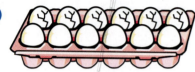

    A How many eggs in the picture are cracked?

    B What fraction, other than $\frac{1}{2}$, describes the cracked eggs in the picture?

---

2 Use a fraction to describe the number of cracked eggs in each group.

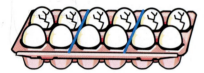

---

3 Use a fraction to describe the number of cracked eggs in each group.

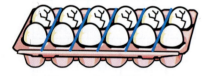

---

4 What is another fraction that describes a half dozen eggs?

---

5 In each picture below, what fraction of the eggs are cracked?

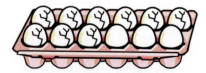 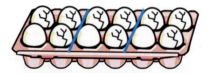

Is the same fraction of eggs cracked in each picture?

108 Chapter 7

Chapter 7 Lesson 5
EXPLORE
Parts of a Dozen

One egg is $\frac{1}{12}$ of a dozen.

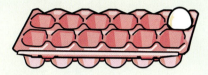

Two eggs are $\frac{2}{12}$ of a dozen.

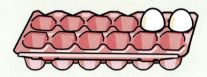

1 What part of a dozen is 5 eggs?

Four eggs are $\frac{4}{12}$ of a dozen.

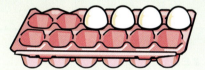

Four eggs are also $\frac{2}{6}$ of a dozen.

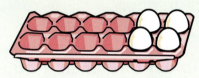

2 Write another fraction that describes 4 eggs as part of a dozen.

3 What part of a dozen is 3 eggs? Write two fractions.

Crystal has $\frac{2}{3}$ of a dozen eggs, and her friend has $\frac{3}{4}$ of a dozen eggs.

4 How many eggs does Crystal have?

5 How many eggs does her friend have?

6 Which is more: $\frac{2}{3}$ or $\frac{3}{4}$?

Chapter 7, Lesson 6

EXPLORE: Fractions of an Hour

Use the clock to help you answer the questions.

1. How many minutes are in an hour?

2. How many minutes are in a half hour?

3. How many minutes are in a quarter of an hour?

Ron will eat dinner in 1 hour. He plans to read for 20 minutes, do homework for 20 minutes, and play for 20 minutes.

4. What fraction of an hour is 20 minutes?

5. What fraction of an hour is 40 minutes?

6. Anh spends $\frac{3}{4}$ of an hour at karate class and $\frac{2}{3}$ of an hour playing piano.

   Which activity lasts longer?

7. Yori walks his dog for $\frac{1}{2}$ of an hour every day. Then they play together for $\frac{1}{3}$ of an hour.

   Which activity lasts longer?

Chapter 7
Lesson 6
REVIEW MODEL
Using Models to Compare Fractions

You can use fraction models to help you compare fractions.

Example Compare $\frac{2}{5}$ and $\frac{3}{8}$.

Compare the lengths of the models for each fraction.

$\frac{2}{5}$ is longer than $\frac{3}{8}$, so $\frac{2}{5} > \frac{3}{8}$.

✓ Check for Understanding

Compare the fractions using <, >, or =.

1. $\frac{1}{2}$ ● $\frac{5}{10}$

2. $\frac{3}{12}$ ● $\frac{3}{8}$

3. $\frac{4}{5}$ ● $\frac{5}{6}$

4. $\frac{2}{3}$ ● $\frac{1}{4}$

5. $\frac{1}{6}$ ● $\frac{1}{3}$

6. $\frac{5}{12}$ ● $\frac{4}{10}$

Chapter 7
Lesson 7
REVIEW MODEL
Problem Solving Strategy
Make a Model

> Kiki bought a dozen muffins. $\frac{2}{3}$ of the muffins were blueberry. $\frac{1}{4}$ of the muffins were cranberry. Did Kiki buy more blueberry or cranberry muffins?

Strategy: Make a Model

Read to Understand

What do you know from reading the problem?

$\frac{2}{3}$ of a dozen muffins are blueberry. $\frac{1}{4}$ of a dozen muffins are cranberry. You need to find out if there are more blueberry or cranberry muffins.

Plan

How can you solve this problem?

You can make a model.

Solve

How can you make a model?

You can use 12 counters to represent the dozen muffins.

To find the number of blueberry muffins, separate the counters into 3 equal groups. There are 8 counters in two groups, so Kiki bought 8 blueberry muffins.

To find the number of cranberry muffins, separate the counters into 4 equal groups. There are 3 counters in one group, so Kiki bought 3 cranberry muffins.

8 > 3, so Kiki bought more blueberry muffins.

Check

Look back at the problem. Did you answer the question that was asked? Does the answer make sense?

Problem Solving Practice

Use the strategy *make a model*.

1. Lupe has some bags of coins. Each bag has 8 coins, and 7 of the 8 coins are pennies. How many pennies does Lupe have in 6 bags?

2. Rex made a square design with 25 tiles. He put a green tile in each corner. He used red tiles to complete the outside border. Then he filled in the center with blue tiles. How many blue tiles did Rex use?

Problem Solving Strategies

- ✔ Act It Out
- ✔ Draw a Picture
- ✔ Guess and Check
- ✔ Look for a Pattern
- ✔ Make a Graph
- ✔ **Make a Model**
- ✔ Make an Organized List
- ✔ Make a Table
- ✔ Solve a Simpler Problem
- ✔ Use Logical Reasoning
- ✔ Work Backward
- ✔ Write a Number Sentence

Mixed Strategy Practice

Use any strategy to solve. Explain.

3. Mr. Ortega is ordering 20 packages of erasers. A package of erasers costs 50¢. If he buys them in bulk, he can get 20 packages for $7.99. How much would he save if he buys in bulk?

4. Carl has 6 library books and borrows 3 more. Mary Beth has 10 library books. If Carl returns 2 books, what must Mary Beth do to have the same amount as Carl?

5. Ms. Holt wrote some fractions. If she continues the pattern, which fraction will she write next?

   $\frac{1}{4}, \frac{2}{8}, \frac{3}{12}, \frac{4}{16}, \underline{\quad ? \quad}$

6. There are 30 people having lunch together. They want to share large submarine sandwiches. Each person will have $\frac{1}{6}$ of a sandwich. How many whole sandwiches should they order?

7. Lester can ride his bike 12 miles in 1 hour. How far can he travel in 6 hours?

8. Trina is going to ride her bike to visit her friends Hank, Marsha, and Annette. How many different ways can she order her visits?

Chapter 7

Chapter 7 Vocabulary

Choose the best vocabulary term from Word List A for each sentence.

1. Two numbers that have the same value are __?__.

2. A fraction with 3 in the denominator describes a group divided equally into __?__.

3. Fifteen minutes is a __?__ hour.

4. A package of twelve items is one __?__ items.

5. A fraction with 4 in the denominator describes a group divided equally into __?__.

6. When two numbers are compared, the one representing more is __?__ the other number.

7. Thirty minutes is a __?__ hour.

8. A(n) __?__ names the same amount as another fraction.

Word List A

denominator
dozen
equal
equivalent fraction
fourths
fraction
greater than
half
halves
hour
less than
minute
numerator
quarter
thirds

Complete each analogy using the best term from Word List B.

9. Two is to __?__ as ten is to tenths.

10. Minute is to __?__ as day is to week.

Word List B

dozen
fraction
halves
hour

Talk Math

Discuss with a partner what you have learned about fractions. Use the vocabulary terms *fraction*, *half*, and *equal*.

11. How can you tell what fraction of a grid is shaded?

12. How can you tell whether two fractions have the same value?

13. How can you use what you know about minutes and hours to compare fractions?

Venn Diagram

14. Create a Venn diagram for the words *hour* and *minute*. Write activities that you would measure by the hour and by the minute.

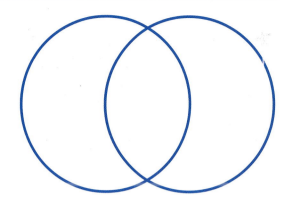

Word Line

15. Create a word line using the words *fourth, half,* and *third.*

Words:

Sequence:

What's in a Word?

HOUR The words *hour* and *our* sound exactly the same, but they have very different meanings. Using the word *our* says two things. One is that the person speaking is part of a group. The other is that the group owns something.

The word *hour,* however, refers only to time. One *hour* is exactly 60 minutes long.

Chapter 7 115

Fraction Construction Zone

Game Purpose
To find equivalent fractions

Materials
- Activity Masters 79–81
- Paper bag
- Scissors

How to Play the Game

1 This is a game for two players.
- Both players will use the Fraction Construction Zone gameboard and cards. Cut out 1 row of the Fraction Construction Zone cards, and put them in the paper bag.
- Each player will need all of the Fraction Pieces. Cut them out.

2 Take turns. Without looking, pick a card from the bag.
- Use the fraction pieces named on the card. If *free choice* is picked, you may choose any size you want.
- Name an empty fraction bar on the gameboard. Use as many fraction pieces as you need to make that bar.
- Write the equivalent fraction. You get 1 point for each fraction piece you used to make the bar.
- If you cannot make the bar you named, you lose your turn and must take away your fraction pieces.

3 Take turns until all the fraction bars have been made. The player with more points wins.

Marble Mystery

Game Purpose
To find fractional parts of sets of objects

Materials
- Activity Master 82
- Two-color counters

How to Play the Game

1. This is a game for 2 players. You each will need a gameboard and some counters. Choose who will be the Mystery Maker and who will be the Detective.

2. The Mystery Maker secretly sets up Game 1. Decide on the number of black marbles in a bag. Shade that number of marbles in each bag. Record the fractions.

3. The Detective asks questions to find the number of black marbles. The questions must be about the total number of marbles. You may NOT ask questions about the marbles in one bag. You may ask a question such as: *Are there more black marbles than white marbles?* You may NOT ask a question such as: *Are there 3 black marbles in a bag?* Use counters to help.

4. The Mystery Maker keeps track of the number of questions. The Mystery Maker gets **1 point** for every question asked. The Detective gets **2 points** for:
   - finding the total number of black marbles.
   - telling how many black marbles are in each bag.
   - modeling the bags with counters.

5. Switch roles. Play Game 1 again. The player with more points after two games wins.

CHALLENGE

Work with a partner to discover how many different ways you can make one whole using halves, fourths, eighths, and sixteenths.

You will need 6 round paper plates, all the same size. You will also need scissors, a ruler, markers, and a bag.

Color 4 paper plates each a different color. Then cut up the paper plates this way:

A Cut one plate in half. Label each part $\frac{1}{2}$.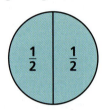

B Cut one plate into 4 equal parts. Label each part $\frac{1}{4}$.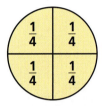

C Cut one plate into 8 equal parts. Label each part $\frac{1}{8}$.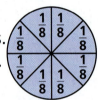

D Cut one plate into 16 equal parts. Label each part $\frac{1}{16}$.

Mix up all the fraction parts and put them in a bag.

Create a New Whole Plate

1. Each partner starts with one whole plate. The goal is to exactly cover up your plate with fraction parts.

2. One partner picks a fraction part from the bag.

3. Take turns. If you get a fraction part and you cannot put it on your plate without overlapping, choose another part.

4. Record the fraction parts you used to cover your plate.

5. Start over. Can you cover the plate another way?

Example: On her first turn, Cho gets $\frac{1}{4}$. After 3 turns, Cho has $\frac{1}{4}$, $\frac{1}{8}$, and $\frac{1}{8}$.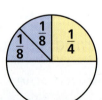

Chapter 8
Charts and Graphs

Dear Student,

The way you display information can help you see patterns and draw conclusions. For example, imagine that your class voted on whether to have recess before or after lunch. Here are the ballots:

Which choice got the most votes? It is difficult to tell until you organize the information:

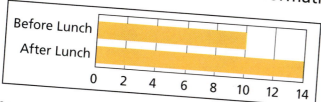

Now it's easier to see which time the class prefers.

In this chapter, you will use tables and graphs to organize information in different ways to help you solve different problems.

Mathematically yours,
The authors of *Think Math!*

THE WORLD ALMANAC FOR KIDS

Wheels in the Air

FACT ACTIVITY 1

How high can you jump? Can you flip and twist in the air? Skateboarders can perform amazing jumps and tricks with just a board and 4 wheels. There are different lengths of skateboards. The following chart shows the average length of some skateboards.

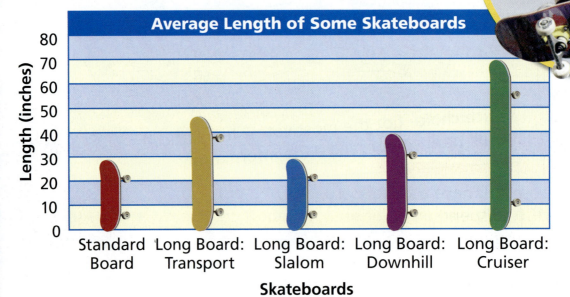

Average Length of Some Skateboards

(Bar chart — Length in inches: Standard Board, Long Board: Transport, Long Board: Slalom, Long Board: Downhill, Long Board: Cruiser)

Use the chart for the problems.

1. How long is the Slalom long board? Explain how you know.
2. How long is the Downhill long board?
3. How much longer is the Cruiser long board than the Standard board?
4. Name two boards whose combined length would be the same as the Cruiser long board.

120 Chapter 8

FACT ACTIVITY 2

The popularity of skateboarding has created a demand for more public skate parks. More cities are providing parks for skateboarders to practice their riding tricks.

Use the map to answer the questions.

1. At what 2 positions would you find the iron railing?
2. What is located at G6?
3. At what 3 positions would you find the ramp?
4. If the park designers wanted to make the ramp longer, at what position could they add an extension?

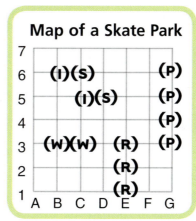

Map of a Skate Park

KEY:
iron railing (I)
stairwell (S)
ramp (R)
pipe (P)
wall (W)

CHAPTER PROJECT

Suppose you have a skateboard, and you have saved up $100 to spend on additional skateboarding gear. Use newspapers, magazines, and catalogs to find the typical price of each of the following items: pads (knee and elbow, pair of each), wrist guards (pair), helmet, wheels (set of 4).

Make a list of the items and their prices. Then determine all the ways you can spend your money without exceeding $100. Make a chart to help you plan your possible purchases. On your chart, include the total cost of what you can buy and your change from $100.

- Can you buy all four of the items? Explain.
- Can you buy three of the items? Which ones?

ALMANAC Fact

Danny Way, a pro skateboarder, performed a stunning jump in the summer of 2005 by jumping over the Great Wall of China from a 9-story "MegaRamp."

Chapter 8
Lesson 3
EXPLORE
Different Pictographs, Same Data

1 Students at Lincoln Elementary were surveyed about school lunch. They were asked to choose their favorite lunch from these five choices. Here are the results:

H	Hamburger or cheeseburger	33
T	Tuna sandwich	24
C	Chicken tenders	72
M	Macaroni and cheese	48
P	Pizza	66

A On a separate sheet of paper, make a pictograph to show the data. Use a plate symbol (O) to represent 6 students.

B How did you represent the results for hamburger or cheeseburger in your pictograph?

2 Here is another pictograph of the same data, where each O represents 10 students. (Each number of students is rounded to the nearest multiple of 10.)

Looking at this pictograph, what is the most popular lunch?

122 Chapter 8

Chapter 8
Lesson 4
EXPLORE
Tossing Two Number Cubes

Imagine that you toss two number cubes and find the sum of the results.

(Each number cube is numbered 1 through 6.)

Classify the following events as *possible (P) or impossible (I)*.

1 The sum is 6.

2 The sum is 4.

3 The sum is 1.

4 The sum is 9, and one cube shows 2.

5 The sum is 8, and one cube shows 4.

6 The sum is 10, and neither cube shows 5.

7 The sum is 13.

8 The sum is multiple of 7.

---

9 Toss two number cubes 30 times and record the sum for each toss.

10 On a separate sheet of paper, make a pictograph of the data from Problem 9 as shown at right.

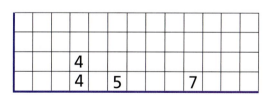

Chapter 8 **123**

Chapter 8 Lesson 4
REVIEW MODEL
Describing the Likelihood of an Event

You can describe the likelihood of an event.

An event is **possible** if it could happen.

Example: Spin a number greater than 2 on the spinner shown. The numbers 4, 6, 8, 10, and 12 are greater than 2, so this could happen.

An event is **impossible** if it can never happen.

Example: Spin a 1. There are no 1s on the spinner, so this can never happen.

More likely than and **less likely** than are used to compare the likelihood of two events.

Example: You are **more likely** to spin a 2 than a 6.

Example: You are **less likely** to spin an 8 than a 4.

✓ Check for Understanding

For **1 to 6**, use the spinner above.

On a separate piece of paper, write *possible* or *impossible* for each event.

1 Spin an even number.

2 Spin an odd number.

3 Spin a number greater than 12.

On a separate piece of paper, write *more likely* or *less likely*.

4 You are ___?___ to spin a one-digit number than a two-digit number.

5 You are ___?___ to spin a number greater than 10 than a number less than 10.

6 You are ___?___ to spin a 4 than a 6.

Chapter 8 Lesson 5
REVIEW MODEL
Listing Outcomes

You can use a table to list possible outcomes for an experiment.

How many possible outcomes are there if you toss two coins?

There is 1 way to get 2 heads.

There are 2 ways to get 1 heads and 1 tails.

There is 1 way to get 2 tails.

So, there are 4 possible ways the two coins can land.

POSSIBLE WAYS FOR TWO COINS TO LAND	
First Coin	Second Coin
heads	heads
heads	tails
tails	heads
tails	tails

✓ Check for Understanding

On a separate piece of paper, complete the table and answer the question.

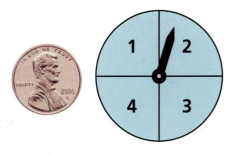

1 How many possible outcomes are there if you toss a coin and spin the pointer shown above?

Coin	Number
heads	1
?	■
?	■
?	4
tails	■
?	■
?	■
?	■

Chapter 8

Lesson 6

EXPLORE
Prices at the Class Store

Erasers and pencils are on sale!

Limit: no more than 3 of each item to a customer.

(Pencil: ~~5¢~~ 4¢; Eraser: ~~4¢~~ 3¢)

1 Chani bought 2 pencils and 2 erasers. How much did she spend?

2 If Chani gave the cashier a quarter, how much change did she receive?

3 List all the purchases you could make for 10¢ or less.

4 Miya spent exactly 17¢. What did she buy?

Chapter 8 Lesson 8
REVIEW MODEL
Using a Map Grid

You can name locations on a map grid.

Activity Jerome's house is at the intersection of two streets. Name the location of Jerome's house.

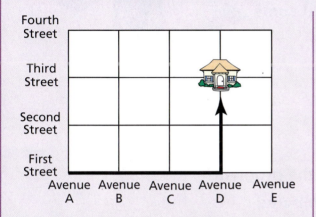

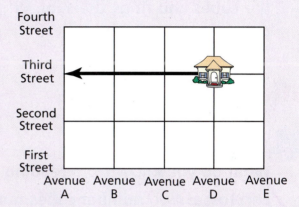

Step ❶ Trace along the bottom of the grid until you reach the vertical line that crosses Jerome's house. Look at the label below the graph that gives the street name.

Step ❷ Find the horizontal line that crosses Jerome's house. Find the label on the side of the grid that shows the name of the other street.

So, Jerome's house is at the intersection of Avenue D and Third Street.

✓ Check for Understanding

Name the location of each house.

❶ Sue's house is at the intersection of __?__ Street and __?__ Avenue.

❷ Tom's house is at the intersection of __?__ Street and __?__ Avenue.

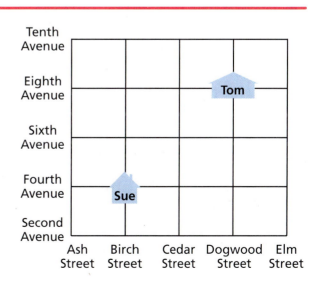

Chapter 8 **127**

Chapter 8
Lesson 10

REVIEW MODEL
Problem Solving Strategy
Make a Table

Lori has two number cubes. Each cube has the numbers 1, 1, 2, 2, 3, and 3. She tosses the cubes and finds the sum. What sums can she toss?

Strategy: Make a Table

Read to Understand

What do you know from reading the problem?

Lori tosses two number cubes numbered 1, 1, 2, 2, 3, and 3 and finds the sum.

What do you need to find out?

All the possible sums Lori can toss.

Plan

How can you solve the problem?

I can make a table.

Solve

How can you make a table to solve the problem?

List all the different tosses for each number cube in the table. Record the sums for each toss in the table. Then look for all the possible sums.

Number Cube 2

Number Cube 1	1	2	3
1	2	3	4
2	3	4	5
3	4	5	6

Lori can toss a sum of 2, 3, 4, 5, and 6.

Check

Look back at the problem. Did you answer the question that was asked? Does the answer make sense?

128 Chapter 8

Problem Solving Practice

Make a table to solve.

1. You have only dimes, nickels, and pennies in your bank. You want to buy a pen for 16¢. What are all the ways you can pay for the pen?

2. Sam would like to buy stickers to decorate his notebook. One sticker costs 12¢. Two stickers cost 24¢, and three stickers cost 36¢. If Sam has a total of 72¢ to spend, how many stickers can he buy?

Problem Solving Strategies

✔ Act It Out
✔ Draw a Picture
✔ Guess and Check
✔ Look for a Pattern
✔ Make a Graph
✔ Make a Model
✔ Make an Organized List
✔ **Make a Table**
✔ Solve a Simpler Problem
✔ Use Logical Reasoning
✔ Work Backward
✔ Write a Number Sentence

Mixed Strategy Practice

Use any strategy to solve. Explain.

3. Annabelle eats 5 servings of vegetables each day. How many servings of vegetables does Annabelle eat in a week?

4. If $\frac{1}{4}$ of a box of crayons is 12 crayons, how many crayons are in the whole box?

5. It takes Manuel 15 minutes to ride his bike to Jake's house. Manuel and Jake want to play video games together for an hour. What time should Manuel leave his house to play video games with Jake and be back home at 6:00 P.M.?

6. Sasha tossed a coin 15 times. She tossed heads twice as many times as tails. How many times did Sasha toss heads?

Chapter 8   129

Chapter 8 Vocabulary

Choose the best vocabulary term from Word List A for each sentence.

Word List A
- data
- impossible
- intersection
- less likely than
- more likely than
- possible
- price
- scale
- spend
- survey
- whole number

1. The number 8 is a(n) __?__, but $3\frac{1}{2}$ is not.
2. Walking to the sun is a(n) __?__ event.
3. A group of questions used to collect data is called a(n) __?__.
4. Going to school on Friday is __?__ going to school on Saturday.
5. A(n) __?__ shows how to measure bars in a bar graph.
6. The place where two lines cross each other is called an __?__.

Complete each analogy. Use the best term from Word List B.

Word List B
- bar graph
- data
- experiment
- possible

7. Symbol is to pictograph as bar is to __?__.
8. Word is to story as __?__ is to graph.

Talk Math

Discuss with a partner what you have learned about pictographs and bar graphs. Use the vocabulary terms *data*, *graph*, *label*, and *symbol*.

9. How can you make a bar graph?
10. How are a pictograph and a bar graph alike? How are they different?

Word Definition Map

11 Create a word definition map for the term *survey*.

A What is it?

B What is it like?

C What are some examples?

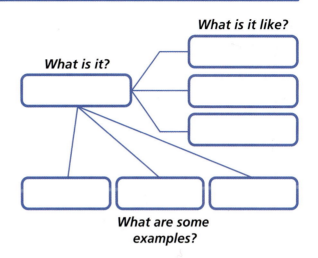

Venn Diagram

12 Create a Venn diagram about pictographs and bar graphs. Use the words *data*, *graph*, *label*, *scale*, and *symbol*.

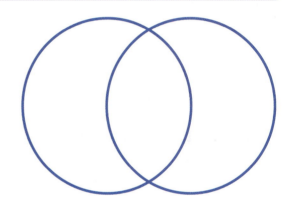

EXPERIMENT An *experiment* is "a trial or a test." *Experiments* are done to discover something, test a guess, or try out a new idea. We use the word *experiment* in everyday life, in science, and in math. School cafeteria workers might do an *experiment* to see if offering more food choices will affect how many lunches are sold. In a science *experiment*, a scientist might test how weather affects plant growth. A math *experiment* could be tossing a coin many times to see how often heads is tossed. For most *experiments*, the results are recorded. Then conclusions can be made from the data.

GO ONLINE Technology
Multimedia Math Glossary
www.harcourtschool.com/thinkmath

GAME

Where's My House?

Game Purpose
To practice locating an object on a grid

Materials
- Activity Master 88: Where's My House?
- Activity Master 89: House Pieces
- Two-color counters

How to Play the Game

1 Play this game with a partner. Each player will need a *Where's My House?* gameboard and 1 house piece.
- Stand an open book or folder between you and your partner so you cannot see each other's gameboard.
- Secretly place your house in one square of your gameboard.

2 Decide who will play first.
- Take turns guessing the location of your partner's house. Ask whether the house is in a certain square.
- Your partner says, "yes," "no," or "near." (Near means your guess is one of the eight squares touching the square with the house.)
- Use two-color counters to mark your guesses. Use one color for "no." Use the other color for "near."

3 The winner is the first player to find the other player's house. Play as many rounds as time allows.

Where's My Car?

Game Purpose
To practice naming intersections on a map grid

Materials
- Activity Master 90: Town Street Map
- Tokens
- Two-color counters

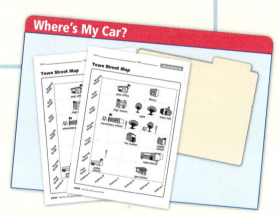

How to Play the Game

1. Play this game with a partner. Each player will need a Town Street Map and 1 token to use as a car.
   - Stand an open book or folder between you and your partner so you cannot see each other's map.
   - Secretly place your car on a street or avenue on your map. The car can be at an intersection or between intersections.

2. Decide who will play first.
   - Take turns guessing the location of your partner's car. Ask whether it is at a certain intersection, between intersections, or anywhere along a certain street.
   - Use two-color counters to mark your partner's responses to your guesses. Use one color for "no." Use the other color for "near."

3. The winner is the first player to find the other player's car. Play as many rounds as time allows.

CHALLENGE

Ozzie surveyed his classmates. He asked "How many days last week did your ride your bicycle?" Ozzie made a table to show the results.

NUMBER OF DAYS LAST WEEK WE RODE OUR BICYCLES	
Number of Days	Number of Students
0	2
1 or 2	4
3 or 4	8
5 or 6	6
I don't ride a bicycle.	4

Make a bar graph or a pictograph from the data in the table. Then use the table or your graph to answer the questions.

1 How many classmates answered the survey question?

2 How many classmates rode their bicycles last week?

3 How many classmates rode more than 2 days?

4 How many classmates who rode their bicycles said they rode fewer than 3 days?

Chapter 9
Exploring Multiplication

Dear Student,

To multiply larger numbers, such as **27 × 58**, it helps to know some multiplication facts well. Some facts that might help you solve this problem are shown below.

Do you already know these facts?

In this chapter, you will improve your knowledge of multiplication facts, and begin to see how these facts can help you multiply larger numbers. The games and puzzles will also give you many chances to show what you already know.

Mathematically yours,
The authors of **Think Math!**

8 × 7

2 × 5

5 × 7

8 × 2

THE WORLD ALMANAC FOR KIDS

Collections

Are you a collector? Some people collect model cars, dolls, or stamps. Others collect more unusual items such as potato chip bags or gum wrappers.

Lauren collects key chains. The table below shows the types of key chains Lauren collects.

FACT·ACTIVITY 1

Use the table to answer the questions.

1. How many types of key chains are in each group? Animals? Fashion? Sports? Vehicles?
2. Lauren has 5 of each type of fashion key chain. How many fashion key chains does she have?
3. Lauren has 9 of each type of animal key chain. How many animal key chains does she have in her collection?
4. Find the total number of sports key chains Lauren has if she has 8 of each.
5. Lauren has 2 of each type of vehicle key chain. Does she have more vehicle or more fashion key chains?
6. Lauren also collects key chains of attractions from U.S. states. She has 5 key chains each from 14 different states. Draw an array to help you find 5 × 14. Break the array into two smaller arrays. What is the total number of state key chains?

Lauren's Key Chain Collection	
Key Chains	**Types**
Animals	cats, birds, fish, dogs
Fashion	shoes, hats, jewelry
Sports	baseball, football, soccer, basketball, tennis, car racing, golf
Vehicles	cars, boats, bikes, motorcycles, planes, trains

baseball

basketball

football

soccer

Sports card collecting started in the 19th century. It became more popular around 1933 when baseball stars such as Babe Ruth and Ty Cobb were featured.

FACT·ACTIVITY 2

Matt has a collection of 120 sports cards that includes baseball, football, basketball, and soccer cards. Use Matt's collection for 1–3.

1. Matt arranged his cards in an array with 12 rows. How many columns are in his array?

2. In Matt's display, there are 4 rectangular sections, one for each sport. Break his array into 4 different sections to show one possibility for Matt's display.

3. Using Matt's array, how many cards of each sport are in each section? Write a multiplication sentence for each section.

CHAPTER PROJECT

Choose an item you would like to collect. Gather, draw, or find pictures of at least 36 of the items.

Separate the collection into at least 3 different groups, either by color, size, or some other attribute. Arrange some or all of the items or pictures in each group in an array. Lay the groups out on a table.

Draw your array on a poster. Show how multiplication can be used to find the number of items in each group and the total number of items displayed in the collection.

ALMANAC Fact

A key chain is not just a simple metal key ring attached to your keys. You can find all kinds of fancy key chains from cartoon characters to flash lights and even the world's smallest calendar.

Chapter 9
Lesson 1
EXPLORE
Exploring Products

How many ways can you make the products?

Find at least two ways to write each number below as the product of two numbers (not including 1). Use tiles or counters to help you.

Example: 20 = 5 × 4 and 20 = 2 × 10

1 24	**2** 30	**3** 45
4 56	**5** 36	**6** 32
7 48	**8** 16	**9** 28
10 18	**11** 42	**12** 54

You may use grid paper or drawings of intersecting lines to help you find the product.

13 5 × 12	**14** 9 × 8	**15** 11 × 11
16 7 × 11	**17** 11 × 12	**18** 6 × 12
19 11 × 10	**20** 9 × 7	**21** 12 × 12

Chapter 9 Lesson 1
REVIEW MODEL
Using Models to Multiply

You can use models to help you multiply.

You can use counters.

Example 3 × 8 = ■

You can make an array with 3 rows and 8 counters in each row. Then find the total number of counters in your array.

There are 24 counters. So, 3 × 8 = 24.

You can use square tiles.

Example 6 × 6 = ■

You can make an array with 6 rows and 6 tiles in each row. Then find the number of tiles in your array.

There are 36 tiles. So, 6 × 6 = 36.

You can draw intersecting lines.

Example 8 × 7 = ■

You can draw 8 horizontal lines and 7 vertical lines. Then find the number of intersections.

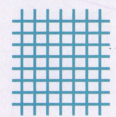

There are 56 intersections. So, 8 × 7 = 56.

You can use grid paper.

Example 5 × 12 = ■

You can shade an array with 5 rows and 12 columns. Then find the number of squares in your array.

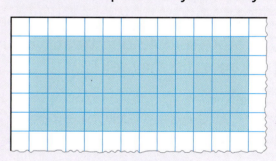

There are 60 squares. So, 5 × 12 = 60.

✓ Check for Understanding

Find the product. You may use any model you wish.

1. 4 × 7 = ■
2. 12 × 6 = ■
3. 9 × 8 = ■
4. 5 × 9 = ■

Chapter 9
Lesson 2
EXPLORE
Missing Factors

Use tiles or counters to solve the problem. Draw a picture to represent your solution, and write a number sentence to describe it.

1 Mrs. Kay gave the same number of stickers to each of her 4 grandchildren. She gave away 36 stickers. How many stickers did each child get?

2 After school, a team of 9 students cleaned up the playground. Each student picked up 4 bags of trash. How many bags did the team collect?

3 The Pet Store sells dog treats in packages of 6 treats. The store sold 48 dog treats yesterday. How many packages did they sell?

4 Cindy is playing a card game to test her memory. She neatly lines up 48 cards in 8 rows and places them face down. How many cards are in each row?

Chapter 9 Lesson 2
REVIEW MODEL
Fact Families

You can write multiplication and division fact families for problem situations.

Example

Angela displays her collection of quarters in an array.

You can use multiplication and division to describe the array.

Use multiplication to tell how many quarters in the array.

Multiply the number of rows by the number of columns, or

$4 \times 6 = 24$ quarters

multiply the number of columns by the number of rows.

$6 \times 4 = 24$ quarters

Use division to tell how many quarters are in each row or column.

Divide the total by the number of rows.

$24 \div 4 = 6$ columns

Divide the total by the number of columns.

$24 \div 6 = 4$ rows

The same numbers are used in all the facts. These related multiplication and division number sentences are called a **fact family**.

✓ Check for Understanding

Write a fact family for each situation.

1 [array of 3 rows × 8 columns of buttons]

2 Suki picked 32 tomatoes. She gave the same number of tomatoes to each of her 4 neighbors.

Chapter 9
Lesson 5
EXPLORE
Using 10 as a Factor

What pattern can help you multiply by 10?

1 You already know these products.

3×10
10×8
4×10
10×7
10×1

Now try to find these products. Then use a calculator to check your answers.

10×12 10×10

16×10 10×23

45×10 87×10

2 Think about the related number sentences for this sentence . . .

$10 \times 36 = 360$

. . . to help you complete the number sentence and write the related number sentences.

$680 = \blacksquare \times 68$

$\blacksquare = \blacksquare \times \blacksquare$

$\blacksquare = \blacksquare \div \blacksquare$

$10 = \blacksquare \div 68$

3 What number must you multiply by 10 to get 370? Write the four related number sentences.

4 Find two numbers with a product of 530, and write a number sentence that uses those numbers.

See if you can find another pair of numbers with a product of 530, and write a multiplication sentence that uses those numbers.

Chapter 9 Lesson 6
REVIEW MODEL
Making Simpler Problems

You can use smaller arrays to help you find the number of squares in a larger array.

Example A You can draw a line to separate a large array into two smaller arrays.

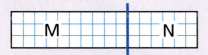

Write a multiplication sentence to find the number of squares in each small array.

M: 3 × 10 = 30   N: 3 × 6 = 18

Add the number of squares in each small array to find the number of squares in the large array.

30 + 18 = 48   So, 3 × 16 = 48

Example B You can draw two lines to separate a large array into four smaller arrays.

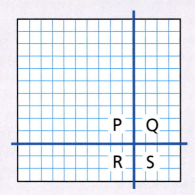

Write a multiplication sentence to find the number of squares in each small array.

P: 10 × 10 = 100   Q: 10 × 4 = 40
R: 3 × 10 = 30   S: 3 × 4 = 12

Add the number of squares in each small array to find the number of squares in the large array.

100 + 40 + 30 + 12 = 182
So, 13 × 14 = 182

✓ Check for Understanding

Find the number of squares in the larger array.

 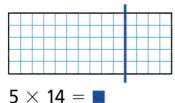

5 × 14 = ■

 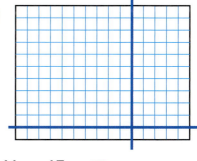

11 × 15 = ■

Chapter 9 Lesson 7
REVIEW MODEL
Problem Solving Strategy
Guess and Check

At the baseball game, Toby spent $6.00 on snacks. He bought a large bag of popcorn and a small bag of peanuts. The bag of popcorn cost 3 times as much as the bag of peanuts. What was the cost of the peanuts? What was the cost of the popcorn?

Strategy: Guess and Check

 Read to Understand

What do you know from reading the problem?

Toby spent $6.00 on a bag of popcorn and a bag of peanuts. The popcorn cost 3 times as much as the peanuts.

 Plan

How can you solve this problem?

You can use the strategy *guess and check.*

 Solve

How can you use this strategy?

Guess the cost of the peanuts, and use this guess to find the cost of the popcorn. Start with an amount less than $6.00, such as $1.00. If peanuts are $1.00, then popcorn is 3 times more, or $3.00: $1.00 + $3.00 = $4.00.

That guess is too low, so adjust the guess. If peanuts are $1.50, then popcorn is $4.50. $1.50 + $4.50 = $6.00. So, that guess is correct.

 Check

Look back at the problem. Did you answer the questions that were asked? Do the answers make sense?

Problem Solving Practice

Use the strategy *guess and check*.

1. The sum of two numbers is 10. Their product is 24. What are the two numbers?

2. There were 125 campers at the cookout. Each camper ordered either a hot dog or a hamburger. There were 25 more hot dog orders than hamburger orders. How many hot dogs were ordered? How many hamburgers were ordered?

Problem Solving Strategies

- ✓ Act It Out
- ✓ Draw a Picture
- ✓ **Guess and Check**
- ✓ Look for a Pattern
- ✓ Make a Graph
- ✓ Make a Model
- ✓ Make an Organized List
- ✓ Make a Table
- ✓ Solve a Simpler Problem
- ✓ Use Logical Reasoning
- ✓ Work Backward
- ✓ Write a Number Sentence

Mixed Strategy Practice

Use any strategy to solve. Explain.

3. Suppose you start an exercise program by exercising 15 minutes a day. If every week you increase your daily exercise time by 5 minutes, during which week would you be exercising 30 minutes a day?

4. At the birthday party, $\frac{5}{6}$ of the chocolate cake got eaten and $\frac{5}{8}$ of the vanilla cake got eaten. If the cakes were the same size, which cake had the greater amount eaten?

5. Six children went apple picking. Each child picked 17 apples. How many apples did they pick?

6. Vincent is watching his favorite movie. The movie is 138 minutes long. If he stops the movie after 2 hours, how many minutes will Vincent have left to watch?

7. Nona is using toothpicks to make the design shown at right. How many toothpicks will Nona need to make 14 squares in a row?

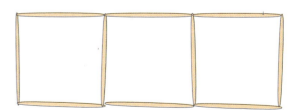

Chapter 9 **145**

Chapter 9 Vocabulary

Choose the best vocabulary term from Word List A for each sentence.

Word List A
array
combine
commutative
Cross Number Puzzle
fact family
factor
product
separate

① Multiplication is __?__ because switching the factors does not change the answer.

② A set of multiplication and division sentences that use the same three numbers is called a(n) __?__.

③ The __?__ is the answer in a multiplication problem.

④ A(n) __?__ is an arrangement of objects in rows and columns.

⑤ You can __?__ rows in two arrays to make a larger array (if the number of columns is the same).

Complete each analogy using the best term from Word List B.

Word List B
fact family
factor
product
array

⑥ Add is to multiply as addend is to __?__.

⑦ Addend is to sum as factor is to __?__.

💬 Talk Math

Discuss with a partner what you have learned about multiplication. Use the vocabulary terms *factor*, *product*, *separate*, and *combine*.

⑧ How can you write a multiplication and division fact family from an array?

⑨ How can you use an array to multiply two numbers?

⑩ How is a Cross Number Puzzle like an array?

Analysis Chart

11. Create an analysis chart for the terms *combine, commutative, fact family, factor,* and *product.*

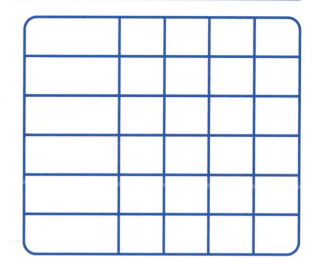

Concept Map

12. Create a concept map using the term *fact family.* Use what you know and what you have learned about addition, subtraction, multiplication, and division.

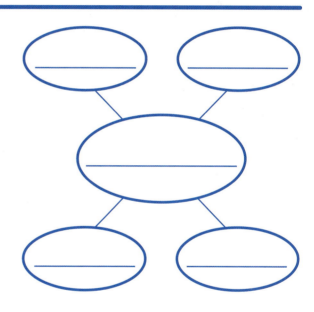

What's in a Word?

FACTOR The word *factor* is often used in everyday life. A *factor* is something that is important for something to happen. It could also be a part of a process. Good study habits are a *factor* in school success. Price could be a *factor* when choosing a new jacket.

In math, the word *factor* also has to do with making something happen. We say 3 and 4 are *factors* of 12 because 3×4 results in 12.

GO ONLINE
Technology
Multimedia Math Glossary
www.harcourtschool.com/thinkmath

GAME

Tic-Tac-Toe Multiplication

Game Purpose
To select factors and find products

Materials
- Activity Masters 95 to 98: Tic-Tac-Toe Multiplication
- Two-color counters

How to Play the Game

1 This is a game for 2 players. You each will need about 20 counters. Decide who will use each color.

2 Place your gameboard (Activity Master 95, 96, 97, or 98) between you. One player faces the numbers in the *Player A* box. The other player faces the numbers in the *Player B* box.

- Player A chooses a number from his or her box. Player A says the number aloud and places a counter on it.
- Player B chooses an unused factor from his or her box that can be multiplied with Player A's factor to make a product on the gameboard. Player B places a counter on the factor and on the product.
- If Player B cannot make a product, he or she loses a turn. Then Player B must name the next factor for Player A.

3 Take turns choosing factors and placing counters on products on the gameboard.

4 The first player to cover 3 products in a line—across, down, or diagonally—wins.

GAME

Caught in the Middle

Game Purpose
To practice identifying factors and products

Materials
• Index cards

How to Play the Game

1 This is a game for 3 players. Make 4 sets of index cards numbered 1 through 12.

2 Mix up the cards. Place them face down in a pile.

- Each player takes two cards and turns them face up. Take turns saying aloud the product of your numbers. Be sure to notice who has the product with the value between the other two products—the middle value, not the highest or lowest.

- Take turns naming a different pair of numbers (not including 1) that also make your product. For example, a player who gets 4 and 8 could name 2 and 16, even though 16 is not a card.

- If you can make your product a different way, place any one of your own cards face down in your "won pile." If you have the product in the middle, take the rest of the face-up cards for your "won pile." If there is no middle value because two players have the same product, the player with the different product can take the remaining cards.

3 Play until there are not enough cards for each player to take 2 cards. The player with the most cards in his or her "won pile" wins.

CHALLENGE

Make a copy of this grid. Then draw arrays for as many of the multiplication facts below as you can. Do not overlap arrays. Make each array a different color. Try to fill as much of the grid as you can. Can you fill the entire grid?

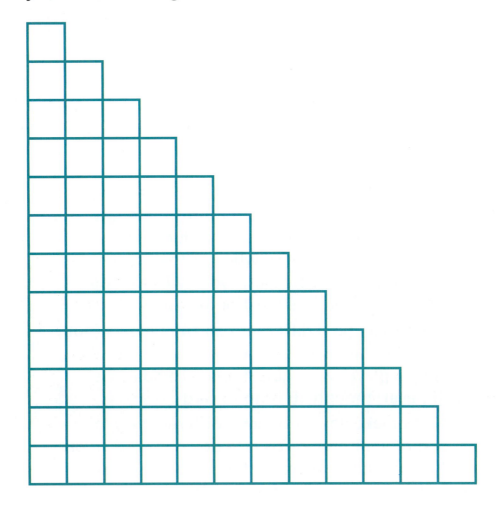

MULTIPLICATION FACTS

1 × 12	1 × 5	1 × 2
2 × 10	1 × 6	1 × 3
3 × 9	2 × 5	2 × 3
4 × 8	3 × 5	3 × 3
5 × 7	5 × 5	4 × 3
6 × 6	4 × 6	6 × 3

Chapter 10
Length, Area, and Volume

Dear Student,

In this chapter, you will be measuring length, area, and volume. These are all measurements of space.

When you use a measuring tape or ruler to measure distance along a straight line or a wiggly path, you are measuring length. When a path loops to make a closed figure like the one shown, you can still measure its length. That length is called the perimeter of the figure. You can also measure something new: the area inside the loop.

Now imagine that you've drawn a loop on the top of a block of wood and cut around it with a saw. You can still measure the length of the loop (perimeter of the figure you've drawn) and the area inside that figure, but the amount of space this lump of wood takes up depends on its thickness. When you measure that amount of space, you are measuring volume.

You'll get a chance to measure the length, area, and volume of many things during this chapter!

Mathematically yours,
The authors of **Think Math!**

Fancy Fish

FACT·ACTIVITY 1

Many people keep tropical fish as pets. Tropical fish are colorful, fun to watch, and they don't need to be taken for a walk!

Goldfish

Royal Gramma Basslet

Betta (Siamese fighting fish)

For 1–3, use a ruler to measure the length of each fish to the nearest quarter inch.

1. Goldfish
2. Betta
3. Royal Gramma Basslet
4. Find an object in the classroom that is about $2\frac{1}{2}$ inches long.
5. Write the lengths of the fish and the object you found in order from least to greatest.

FACT·ACTIVITY 2

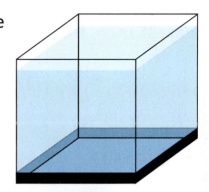

Have you ever wondered how many fish can live together in a tank? This actually depends on the type of fish. A Betta is an aggressive fish and is usually kept alone in a 10-gallon tank or bigger. A Betta can live with other fish in the same tank as long as the other fish are peaceful.

For 1–4, you may use cubes to help.

1. Suppose you have a Betta fish, and you put it in a tank that is 2 feet wide and 2 feet long. What is the perimeter of the tank's bottom?

2. Suppose your tank is 2 feet wide, 2 feet long, and 2 feet tall. Explain how you can find the volume of the tank in cubic feet.

3. Suppose you want to keep a goldfish and a Basslet together with your Betta in a larger tank. The new tank is 4 feet wide, 5 feet long, and 3 feet high. Find the volume of the tank.

Hint: Remember that one cube represents one cubic foot for these problems.

CHAPTER PROJECT

Materials: cubes

Design your own large fish tank. Your tank needs to have a volume of 48 cubic feet.

- Use cubes to build 3 different rectangular fish tank designs.
- Find the perimeter of your tank's bottom.
- Find the area of your tank's bottom.
- If the bottom of your fish tank has an area of 4 square feet, what is the tallest your tank could be? Explain how you can use cubes to find out.
- Describe how you could find the total area of the glass sides of any of your tanks. Do not include the top.

ALMANAC Fact

Americans own about 50 million pet fish, making fish the third most popular pet after cats and dogs.

Chapter 10
Lesson 1

REVIEW MODEL
Measuring to the Nearest Inch, $\frac{1}{2}$ Inch, and $\frac{1}{4}$ Inch

You can use a ruler to measure objects to the nearest inch, half inch, and quarter inch.

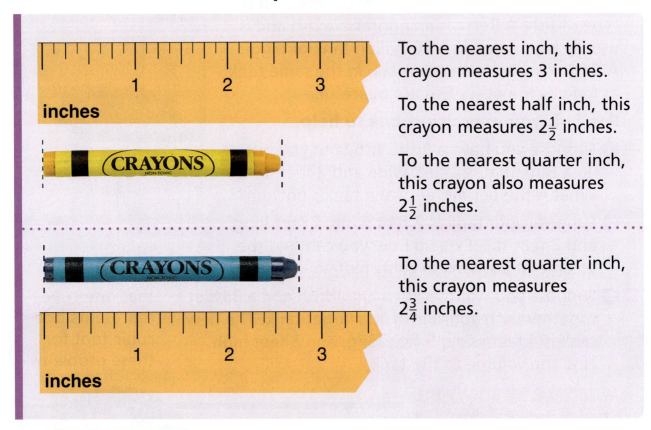

To the nearest inch, this crayon measures 3 inches.

To the nearest half inch, this crayon measures $2\frac{1}{2}$ inches.

To the nearest quarter inch, this crayon also measures $2\frac{1}{2}$ inches.

To the nearest quarter inch, this crayon measures $2\frac{3}{4}$ inches.

✓ Check for Understanding

1 What is the measure of the ribbon to the nearest half inch?

2 What is the measure of the ribbon to the nearest quarter inch?

Chapter 10
Lesson 5
EXPLORE
Measuring Area and Perimeter

Each small square on this page measures **1 cm** on each of its four sides.

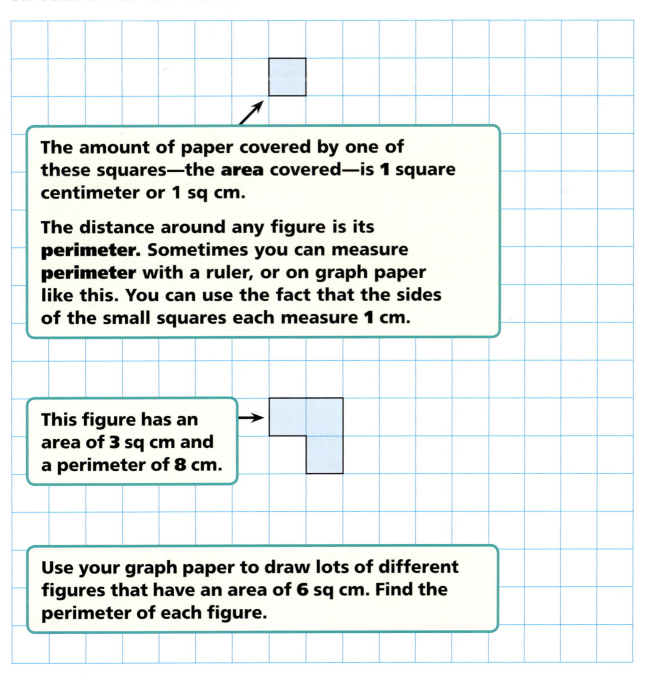

The amount of paper covered by one of these squares—the **area** covered—is **1** square centimeter or **1 sq cm**.

The distance around any figure is its **perimeter**. Sometimes you can measure **perimeter** with a ruler, or on graph paper like this. You can use the fact that the sides of the small squares each measure **1 cm**.

This figure has an area of **3 sq cm** and a perimeter of **8 cm**.

Use your graph paper to draw lots of different figures that have an area of **6 sq cm**. Find the perimeter of each figure.

Do all the figures have the same perimeter? If not, find as many different perimeters as you can.

Chapter 10
Lesson 5
REVIEW MODEL
Measuring Perimeter and Area

Perimeter (P) is the distance around a figure. Area (A) is the number of square units needed to cover a flat surface.

You can measure the perimeter of a figure in centimeters (cm) and the area of a figure in square centimeters (sq cm).

The perimeter of the figure is 10 cm.

The area of the figure is 4 sq cm.

Two figures with the same area can have different perimeters.

P = 8 cm
A = 4 sq cm

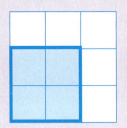

P = 10 cm
A = 4 sq cm

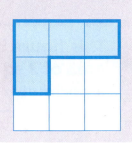

✓ Check for Understanding

1 Which two figures have the same perimeter?

2 Which two figures have the same area?

156   Chapter 10

Chapter 10
Lesson 6 — EXPLORE: An Ant Corral

> Imagine that you are an ant wrangler, and you're making a tiny corral for your herd of ants.
>
>
>
> You have 16 cm of fencing for the corral.

Use a piece of centimeter graph paper to try some designs. Make the fence follow the lines of the graph paper. Use all 16 cm of fence for each design.

1. Record the area and perimeter of each design.

2. What is the smallest area of the corrals you designed?

3. What is the largest area of the corrals you designed?

4. If you had 18 cm of fencing instead of 16 cm, could you enclose a larger area? What would be the largest possible area?

Chapter 10
Lesson 7
EXPLORE
Building with Cubes

1. Using units from a set of base-ten blocks, build this box:

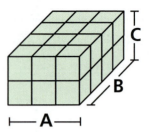

2. Use a centimeter ruler to measure your box.
   - **A** From left to right
   - **B** From front to back
   - **C** From top to bottom

3. How many units did you use to build the box?

4. Using rods from a set of base-ten blocks, build this box:

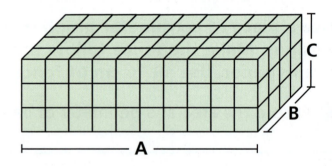

5. Use a centimeter ruler to measure your box.
   - **A** From left to right
   - **B** From front to back
   - **C** From top to bottom

6. Although you actually used rods to build this box, if you built it with units instead, how many units would you need?

Chapter 10 Lesson 7

REVIEW MODEL
Measuring Volume

Volume is the amount of space a three-dimensional figure takes up. This is 1 cubic unit. It is used to measure volume.

You can find volume by counting the number of cubic units needed to fill an object.

Volume = 9 cubic units

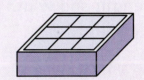

You can find volume by counting the number of cubes in each layer of a three-dimensional figure.

2 layers × 4 cubes in each layer
Volume = 8 cubic units

You can find volume of a rectangular box by multiplying the length, width, and height.

3 cubes × 2 cubes × 4 cubes
Volume = 24 cubic units

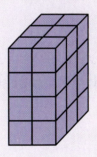

✓ Check for Understanding

Write the volume for each figure in cubic units.

1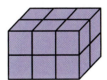

■ cubic units

2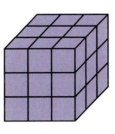

■ cubic units

3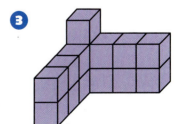

■ cubic units

Chapter 10 **159**

Chapter 10 Lesson 8

REVIEW MODEL
Problem Solving Strategy
Draw a Picture

Katie drew a rectangular figure with an area of 20 sq cm and a perimeter of 24 cm. What are the length and width of the figure she drew?

Strategy: Draw a Picture

Read to Understand

What do you know from reading the problem?

The rectangular figure has an area of 20 sq cm and a perimeter of 24 cm.

Plan

How can you solve the problem?

You can draw a picture to solve the problem.

Solve

How can you draw a picture to solve the problem?

You can draw different rectangles that have an area of 20 sq cm. Then you can find the one that has a perimeter of 24 cm. Find the length and width of this figure.

Check

Look back at the problem. Did you answer the question that was asked? Does the answer make sense?

Problem Solving Practice

Draw a picture to solve.

1. Francisco drew a rectangle with a perimeter of 20 cm. The length of the rectangle is 9 cm. What is the area of the rectangle?

2. Celine is setting up for the book fair. She has 28 books to display on a table. If she places the same number of books in each row, how many different ways could she arrange the books?

Problem Solving Strategies
- ✓ Act It Out
- ✓ **Draw a Picture**
- ✓ Guess and Check
- ✓ Look for a Pattern
- ✓ Make a Graph
- ✓ Make a Model
- ✓ Make an Organized List
- ✓ Make a Table
- ✓ Solve a Simpler Problem
- ✓ Use Logical Reasoning
- ✓ Work Backward
- ✓ Write a Number Sentence

Mixed Strategy Practice

Use any strategy to solve. Explain.

3. Erika has 140 pictures in her photo albums. Her first album has 20 more pictures in it than her second album. How many pictures are in each album?

4. Jackie's karate class begins at 4:45. It takes her 10 minutes to get from home to her class. She spends 20 minutes getting ready for the class. At what time should Jackie begin getting ready for karate?

5. Jared wrote the numbers 3, 6, 8, 11, 13, 16, and 18 on the board. What are the next two numbers in his pattern?

6. Luis is studying for the Spelling Bee. He studied 12 words each night for the last 5 nights. How many words has he studied so far?

7. Nathan is taller than Julio. Emily is shorter than Julio. What is the order of the children from shortest to tallest?

8. Alely used quarters and nickels to pay 80¢ for her snack. She used 2 quarters. How many nickels did she use?

Chapter 10 **161**

Chapter 10 Vocabulary

Choose the best vocabulary term from Word List A for each sentence.

Word List A
- area
- centimeter
- cubic centimeters
- half inch
- height
- length
- perimeter
- quarter inch
- volume
- width

1. The middle mark between 0 and 1 inches measures a(n) __?__.

2. The dimensions of a box are __?__, __?__, and __?__.

3. The __?__ is the distance around something.

4. The __?__ of this figure is the number of squares that can cover it.

5. You can use __?__ to measure volume.

Complete each analogy. Use the best term from Word List B.

Word List B
- area
- centimeter
- cubic centimeter
- perimeter

6. Inch is to foot as __?__ is to meter.

7. Length is to centimeter as volume is to __?__.

Talk Math

Discuss with a partner what you have learned about perimeter, area, and volume. Use the vocabulary terms *length*, *width*, and *height*.

8. How can you measure the perimeter of a sheet of paper?

9. How can you measure the area of a sheet of paper?

10. How can you measure the volume of a box?

Venn Diagram

11 Create a Venn diagram for the terms *height, length, width, area, perimeter, volume, square centimeter, cubic centimeter, square inch,* and *cubic inch.* Call one circle of the Venn diagram "Measurements of a Square." Call the other circle "Measurements of a Box."

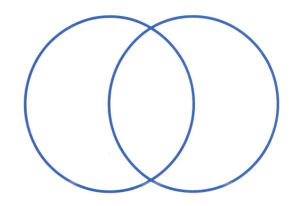

Word Definition Map

12 Create a word definition map for the term *volume.*

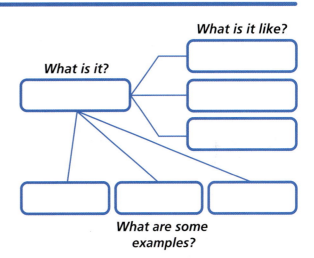

RANGE The word *range* in the song *Home on the Range* means a wide-open area where animals can roam and feed. Another type of *range* is a cooking stove with burners on top and an oven. Both of those meanings of *ranges* are nouns. *Range* can also be a verb. "To *range* through the park" means "to walk around and explore the park."

In mathematics, *range* is a noun. It is the difference between the largest number and smallest number in a set of numbers.

GO ONLINE Technology
Multimedia Math Glossary
www.harcourtschool.com/thinkmath

Ruler Game

Game Purpose
To practice identifying fractional parts of inches

Materials
- Activity Master 104: Ruler Game
- Paper clip and pencil
- Crayons or markers

How to Play the Game

1. Play this game with a partner. To use the spinner on Activity Master 104, lay a paper clip flat. Put the pencil point through one end of it and onto the dot in the center of the spinner. The paper clip will spin around the pencil point.

2. Spin the spinner. The player who spins the greater fraction goes first. That player will choose a ruler on Activity Master 104.

3. The first player spins the spinner and shades his or her ruler for the fraction of an inch shown by the spin.

4. The other player spins and shades his or her ruler the fraction of an inch shown by the spin.

5. Take turns. On each turn, add the fraction of an inch shown by your spin to your shaded section.

    Example: On your first two turns, you spin $\frac{3}{4}$ inch and $\frac{1}{2}$ inch. So, your ruler looks like this:

6. Play until one player gets to 5 inches or more on the ruler. That player wins.

GAME

Perimeter Golf

Game Purpose
To draw figures with a given area and the smallest perimeter possible

Materials
- 2 number cubes, colored pencils or markers
- Activity Master 111: Centimeter Grid Paper

How to Play the Game

1. Play this game with a partner. Each player tosses the number cubes. The player with the larger sum goes first.

2. The first player tosses the 2 number cubes and finds the sum. The player must draw a figure on the grid paper that has an area in square centimeters equal to their sum. The player tries to get the smallest perimeter possible.

3. The player's score for the "hole" is the perimeter of the figure drawn.

4. Take turns tossing the number cubes and drawing a figure. The player with the lower score after 18 holes of *Perimeter Golf* is the winner.

CHALLENGE

You can build models of cubes using smaller cubes.

❶ A cube has 6 faces. If you paint the outside of 1 cube blue, all 6 faces will be blue.

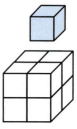

  A How many small cubes do you need to build the next larger cube?

❷ Suppose you paint the outside of the next larger cube blue and take it apart.

  A How many cubes will have 0 blue faces?

  B How many cubes will have 1 blue face?

  C How many cubes will have 2 blue faces?

  D How many cubes will have 3 blue faces?

  E How many small cubes do you need to build the next larger cube?

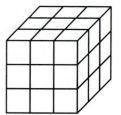

❸ Suppose you paint the outside of the next larger cube blue and take it apart.

  A How many cubes will have 0 blue faces?

  B How many cubes will have 1 blue face?

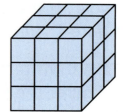

  C How many cubes will have 2 blue faces?

  D How many cubes will have 3 blue faces?

Chapter 11 Geometry

Dear Student,

Think of all the kinds of figures you have seen! Some figures are two-dimensional.

You can find ways in which some of these figures are alike.

- Which figures have only straight sides?
- Which figures have exactly 6 straight sides?
- Which figures have right angles?

Some figures are three-dimensional.

In this chapter, you will learn ways to describe figures using attributes (they tell in which ways figures— and other things—are alike or different). You will also learn the names of some figures and explore characteristics of figures.

Mathematically yours,
The authors of **Think Math!**

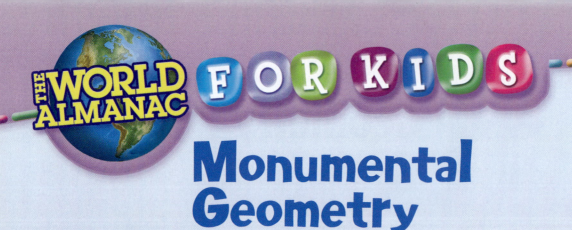

Monumental Geometry

Have you ever visited or seen photos of our nation's capital, Washington, D.C.? Many buildings there honor past presidents of the United States. The Thomas Jefferson Memorial is a tribute to our third president, Thomas Jefferson, who was also the author of The Declaration of Independence.

FACT·ACTIVITY 1

Use the picture below. Answer the following questions about the figures outlined in red.

1. Draw each figure outlined in red. Name the figures.
2. Draw a line of symmetry for each figure. Which of the outlined figures has more than one line of symmetry?
3. Which figures have parallel sides?
4. Which figure has two pairs of parallel sides?
5. Which figure has a right angle?
6. Which figures have no right angles?

Jefferson Memorial

168 Chapter 11

Visitors to Washington, D.C., can also see the Washington Monument, and the Lincoln Memorial. A modern building houses the National Gallery of Art.

FACT·ACTIVITY 2

Look for three-dimensional figures in the photos. Write the name of the photo to answer Problems 1 and 2.

1. Which photos have figures that represent pyramids? Describe where the pyramids appear.
2. Which photo shows figures that represent cylinders? Describe where the cylinders appear.
3. How would you describe the three-dimensional figures of the National Gallery of Art building?

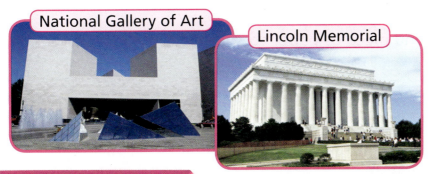

National Gallery of Art

Lincoln Memorial

Washington Monument

CHAPTER PROJECT

Think about historical places in your community. Consider the first library, first post office, or the first school. Research a historical landmark. Using blocks or other materials, create a model of the landmark. Observe the structure from different sides. Name the three-dimensional figures in the structure. Describe the two-dimensional figures on each side. In your description, use the terms *faces*, *edges*, *vertices*, *sides*, and *congruent*.

Chapter 11
Lesson 2 — EXPLORE: Exploring Parallel Sides

Two sides of a quadrilateral are parallel. Are the other two sides also parallel? Complete Steps 1 through 6 to find out what affects the answer.

Step 1
Tape a straw to a large piece of paper so that it is slanted: **not** horizontal (like this: —) or vertical (like this: |).

Step 2
Tape another straw that is the same length to the paper so that it is **parallel** to the first straw. (They form lines that never cross and are always the same distance apart.) Leave lots of room between the straws.

Step 3
Use a ruler to draw 2 straight lines that connect the ends of the straws. This makes a closed figure.

Step 4
Tape a straw to a second piece of paper so it is **not** horizontal or vertical.

Step 5
Tape a straw that is a **different** length to the paper so it is parallel to the first straw.

Step 6
Create another figure with **4 straight sides** by drawing 2 straight lines to connect the ends of the straws.

Compare the two figures you made.
Are the sides you drew for each figure parallel?

Chapter 11, Lesson 2
REVIEW MODEL
Identifying Parallel Lines

You can decide if a pair of lines is parallel. Look to see if the two lines will never cross or if they are always the same distance apart.

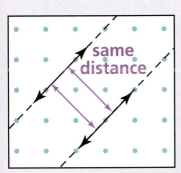

If I extend the lines will they ever cross?

Are the lines always the same distance apart?

The lines will never cross, and they are always the same distance apart. So, the lines are **parallel**.

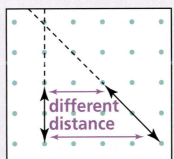

If I extend the lines will they ever cross?

Are the lines always the same distance apart?

The lines will cross, and they are not always the same distance apart. So, the lines are **not parallel**.

Check for Understanding

Write *parallel* or *not parallel* to describe the pair of lines.

1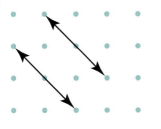

2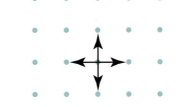

3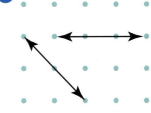

4 Draw a pair of parallel lines. You may use a ruler to help you.

5 Draw a pair of lines that are **NOT** parallel. You may use a ruler to help you.

Chapter 11
Lesson 5
EXPLORE
Exploring Polygons

Part 1 What figures can you make from two triangles?

A On a rectangular card, use a ruler to draw a diagonal line from one corner to another like this:

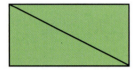

B Cut the rectangle along the line you drew. Check to make sure the two triangles are congruent.

C See how many different figures you can make by placing these two parts next to each other with two congruent sides matching exactly. Trace each figure you make on a separate piece of paper.

Part 2 What figures can you make from a trapezoid and a triangle?

A Fold a rectangular card to divide its long sides exactly in half as shown by the dashed line below. Use a ruler to draw a line from the end of the fold to the corner like this:

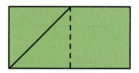

B Cut the card along the line you drew. You should have a triangle and a trapezoid.

C See how many different figures you can make placing these two parts next to each other with two congruent sides matching exactly. Trace each figure you make on a separate piece of paper.

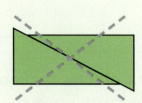

NOT OK
congruent sides touching, but not matching

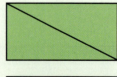
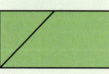

OK
congruent sides matching

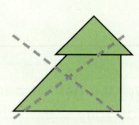

NOT OK
non-congruent sides touching

Chapter 11 Lesson 5
REVIEW MODEL
Sorting Polygons

You can sort figures by different attributes, such as the number of sides, the size of the angles, and the number of pairs of parallel sides.

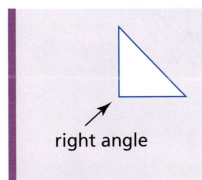

This **triangle** has 3 sides and 1 right angle.

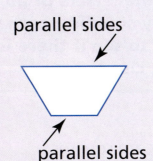

This **quadrilateral** has 4 sides, 1 pair of parallel sides, and no right angles.

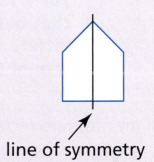

This **pentagon** has 5 sides and 1 line of symmetry.

✓ Check for Understanding

List the figures that belong in each group.

❶ triangles with no right angles

❷ quadrilaterals with at least 1 pair of parallel sides

❸ pentagons with at least 1 line of symmetry

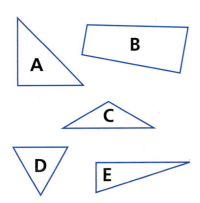

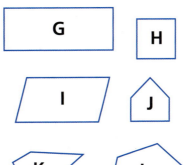

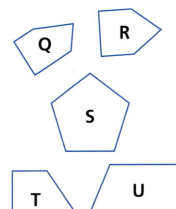

Chapter 11 **173**

Chapter 11
Lesson 8

EXPLORE
Going on a Figure Safari

Which figures from the class Figure Zoo match the clues?

> For 1 to 7, list the letters of all the figures that belong. Try standing each figure on different faces to see if there is any way it might fit the clues.

1 All my faces are rectangles.
Only two of those rectangles are squares.

_____

2 At least three of my faces are triangles.

_____

3 Two of my faces are parallel and congruent to each other.
All my other faces are parallelograms.

_____

4 All my faces are congruent.

_____

5 Two of my faces are parallel and congruent to each other.
All my other faces are not parallelograms.

_____

6 All my faces have at least two lines of symmetry.

_____

7 My top and bottom faces are congruent.
All my other faces are rectangles.

Chapter 11 Lesson 9

REVIEW MODEL
Comparing Three-Dimensional Figures

You can compare three-dimensional figures by describing the faces, edges, and vertices.

Example A: Pyramid

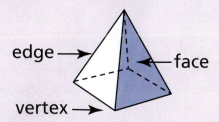

Step ❶

Write the names of the two-dimensional figures for the faces and count the number of each kind of face.

Faces: 1 square and 4 triangles

Step ❷

Count the number of edges and vertices.

Edges: __8__ Vertices: __5__

Example B: Prism

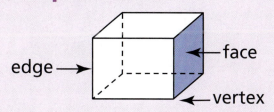

Step ❶

Write the names of the two-dimensional figures for the faces and count the number of each kind of face.

Faces: 6 rectangles

Step ❷

Count the number of edges and vertices.

Edges: __12__ Vertices: __8__

✓ Check for Understanding

Describe each three-dimensional figure. Identify the faces and the number of edges and vertices.

❶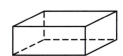

❷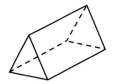

Chapter 11 Lesson 10

REVIEW MODEL
Problem Solving Strategy
Look for a Pattern

Troy sorted some three-dimensional figures into two groups.

Group 1

Group 2

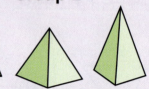

Into which group should Troy place this figure?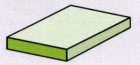

Strategy: Look for a Pattern

Read to Understand

What do you need to find out?

Where Troy should place the next figure

Plan

How can you solve the problem?

You can look for a pattern.

Solve

How can you look for a pattern to solve the problem?

Look to see how the figures in each group are alike. The figures in group 1 are prisms. The figures in group 2 are pyramids. The next figure is a prism, so it should be in group 1.

Check

Look back at the problem. Did you answer the question that was asked?

Problem Solving Practice

Look for a pattern to solve.

1 Anita wants to find out the date of her friend's birthday. She knows it is the fifth Saturday in December. She knows that the dates of the first three Saturdays in December are 2, 9, and 16. What is the date of her friend's birthday?

2 Yoshi is making a pattern. G ↶ ↷ ↶ G

If he continues the pattern in the same way, what will the next figure be?

Problem Solving Strategies

- ✔ Act It Out
- ✔ Draw a Picture
- ✔ Guess and Check
- ✓ **Look for a Pattern**
- ✔ Make a Graph
- ✔ Make a Model
- ✔ Make an Organized List
- ✔ Make a Table
- ✔ Solve a Simpler Problem
- ✔ Use Logical Reasoning
- ✔ Work Backward
- ✔ Write a Number Sentence

Mixed Strategy Practice

Use any strategy to solve. Explain.

3 The length of one side of a rectangle is 2 centimeters. Another side of the same rectangle is twice as long. What is the perimeter of the rectangle?

4 Courtney saves $0.75 the first week, $1.25 the next week, and $1.75 the third week. If this pattern continues, how much money will she save in the fourth week?

5 The baseball team scored 2 runs in the first inning. After 3 innings, the team scored a total of 6 runs. What are all the ways the team could have scored 6 runs after 3 times at bat?

6 Last month, 4 students received an award for perfect attendance. This month, 3 times as many students received a perfect attendance award. How many more students received an attendance award this month than last month?

Chapter 11 Vocabulary

Choose the best vocabulary term from Word List A for each sentence.

1. Two lines that form right angles are __?__ to each other.

2. Three edges of a three-dimensional figure meet at a(n) __?__.

3. A(n) __?__ is a special type of rectangle.

4. A(n) __?__ has three sides.

5. You can fold a pentagon in half in five different ways and have both halves match exactly. This pentagon has five lines of __?__.

6. A(n) __?__ has two faces that are triangles and three faces that are rectangles.

7. A square corner is a(n) __?__.

8. If two figures match exactly, then they are __?__.

9. A two-dimensional pattern for a cube is called a(n) __?__.

Word List A

congruent
flip
net
parallelogram
pentagon
perpendicular
polygon
quadrilateral
right angle
square
symmetry
triangle
triangular prism
turn
vertex

Complete each analogy. Use the best term from Word List B.

10. Triangle is to pyramid as __?__ is to prism.

11. Poodle is to dog as rectangle is to __?__.

Word List B

face
net
quadrilateral
rectangle
triangle

Talk Math

Discuss with a partner what you have learned about geometry. Use the vocabulary terms *side*, *right angle*, *parallel*, and *perpendicular*.

12. How does a rectangle compare to a trapezoid?

13. How can you tell when two rectangles are congruent?

Degrees of Meaning Grid

14. Create a degrees of meaning grid for the words *pentagon, polygon, quadrilateral, rectangle, square, trapezoid,* and *triangle.*

General	Less General	Specific	More Specific

Word Line

15. Use what you know and what you have learned about geometry to create a word line for polygons. Use the number of sides from least to greatest for the sequence.

Words:

Sequence:
_____

RECTANGLE The word *rectangle* comes from an old Latin word that means "having a right angle." Today, *rectangle* means any quadrilateral with four right angles. A square is a special type of *rectangle*.

Technology
Multimedia Math Glossary
www.harcourtschool.com/thinkmath

Chapter 11

GAME

What's My Rule?

Game Purpose
To practice identifying the attributes of two-dimensional figures

Materials
- Activity Masters 113–114: Sorting Figures
- scissors

How To Play The Game

1 Play the game with a small group. Sit around a table or desk. Cut out all the figures.

2 Take turns being the Rule Maker. The Rule Maker secretly makes a rule for sorting the figures into two groups. One group of figures will follow the rule. The other group of figures will not.

Possible rules:
- 3 sides only
- No right angles
- Not 4 sides
- At least 1 right angle

3 The Rule Maker puts the figures on the table one at a time. The first figure follows the rule. After that, he or she sorts the figures using the secret rule.

4 The winner is the first player who correctly names the rule the Rule Maker is using.

Example: There are 2 groups of figures on the table.

Can you guess the rule? If you're the first player to say "No right angles," you win!

5 Play again. Keep playing until everyone has had a chance to be the Rule Maker.

180 Chapter 11

Polygon Bingo

Game Purpose
To practice matching figures with their descriptions

Materials
- Activity Master 119: Bingo Attributes
- Activity Masters 120–123: Polygon Bingo
- Counters, scissors

How To Play The Game

1 Play this game in a small group. Cut out the description cards from Activity Master 119. Place them face down. Each player will need a *Polygon Bingo* board and 20 counters.

2 Take turns picking a description card and reading it aloud.
- If a player has a figure on his or her board that matches the description, the player covers it with a counter.
- There might be more than one figure that matches the description, but you can cover only one figure for each description.

3 If you cover 5 figures in a row, a column, or a diagonal, say "Bingo!" Show your board. Does everyone agree that the winning figures match the descriptions that were read? If yes, you win! If no, keep playing until someone has "Bingo."

4 Clear the counters off the boards for each new game. Players may wish to trade Bingo boards.

CHALLENGE

The popular puzzle below is known as a tangram. Trace the puzzle. Then cut along the lines. There will be 7 pieces in all.

Now use the puzzle pieces to make all the squares you can. You know you can make a square using 1 piece, because 1 piece is a small square. You know you can make a square using all 7 pieces, because that is the shape of the puzzle. Can you make a square using 2 pieces? 3 pieces? 4 pieces? 5 pieces? 6 pieces?

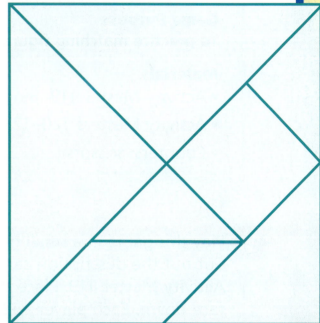

Copy and complete this table. Try to make each square. Good luck!

Can you make a square with	Yes or No	Draw the square if you can make it.
2 pieces?		
3 pieces?		
4 pieces?		
5 pieces?		
6 pieces?		

Chapter 12
Multiplication Strategies

Dear Student,

Multiplying larger numbers is easier when you break the numbers into smaller parts. For example, when finding 14×6, you can think of breaking 14 into 10 and 4.

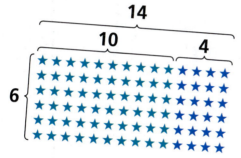

Then you can multiply each part by 6.

$14 \times 6 = 10 \times 6 + 4 \times 6$

Could you break 14 differently to multiply 14 by 6? Could you use the same idea to multiply 27 by 13?

In this chapter, you will be breaking numbers in different ways to make it easier to multiply large numbers.

Mathematically yours,
The authors of *Think Math!*

At the Post Office

Have you ever noticed the many designs on U.S. postage stamps? You can buy stamps with pictures of superheroes, animals, and historic images.

FACT·ACTIVITY 1

Answer the following questions.

1 Marla needs to place 39¢ worth of stamps on an envelope. How can she use 10¢ and 1¢ stamps to equal the value of one 39¢ stamp?

■ × 10¢ = ■ ¢   ■ × 1¢ = ■ ¢

■ ¢ + ■ ¢ = 39¢

2 How much will Marla pay for two 39¢ stamps? Show two ways she can use 10¢ and 1¢ stamps to equal the value of two 39¢ stamps.

■ ¢ + ■ ¢ = ■ ¢

■ ¢ + ■ ¢ = ■ ¢

3 How much would four 39¢ stamps cost? Complete the number sentence to help find the answer.
4 × 39 = (4 × ■) + (4 × ■) = ■

coil

book

pane

You can buy stamps in books of 20 stamps, in panes of 11, 16, or 20 stamps, or in coils of 100 stamps.

FACT ACTIVITY 2

For 1–3, use the chart.

1. How many animal stamps are in 6 panes?
2. Find the total cost for each pane of stamps: Muppet stamps; animal stamps; and butterfly stamps.
3. Ryan uses 3 of the same stamps to mail a package. The total postage is $1.17. Which type of stamp does he use? Write a sentence with a missing factor to solve.

CHAPTER PROJECT

Decide on a shape for a stamp you design. What picture or design will be on the stamp?

Give your stamp a value less than $1.00. How many will be in each book, each pane, and each coil? Find the cost of one book, one pane, and one coil of your new stamp. Use base-ten blocks if you need help with the multiplication.

Present your stamp design and data on a poster. Include a price chart showing the cost of 1 to 3 books, panes, or coils of your new stamp.

The Value of Stamps		
Stamps	Number per Pane	Price per Stamp
Kermit	11	37¢
Curious George	16	39¢
Common Buckeye	20	24¢

Chapter 12
Lesson 1
EXPLORE
Multiplying Money

Use real or play coins to multiply the amount. Then write a multiplication sentence.

1   × 4

2   × 2

3   × 3

4   × 6

5 Make at least two problems using your own combinations of dimes, nickels, and pennies. You do not need to use all three kinds of coins each time.

Chapter 12
Lesson 3
EXPLORE
Multiplying Blocks

Each picture of base-ten blocks represents a number. Find that number, solve the multiplication problem, and write the matching number sentence. You may use blocks to help you.

1 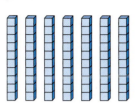 × 4

2 × 5

3 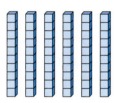 × 5

4 × 4

5 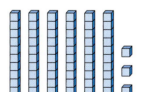 × 5

6 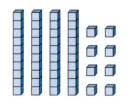 × 5

Chapter 12, Lesson 4

REVIEW MODEL
Using Models to Find Larger Products

You can find larger products by adding the products of two simpler multiplication problems.

Example Find 13 × 4.

One Way

Use base-ten blocks.

× 4

Step 1 Multiply the tens.

10 × 4 = **40**

Step 2 Multiply the ones.

3 × 4 = **12**

Step 3 Add the two smaller products to find the larger product.

40 + **12** = **52**   So, 13 × 4 = **52**.

Another Way

Draw intersecting lines.

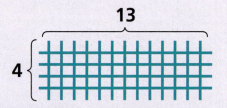

Step 1 Separate the array into two simpler problems. (Breaking a number apart into a multiple of 10 and leftover ones can make it easier to find a solution.)

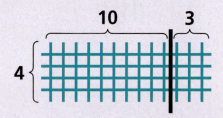

10 × 4 = **40**   3 × 4 = **12**

Step 2 Add the two smaller products to find the larger product.

40 + **12** = **52**   So, 13 × 4 = **52**.

✓ Check for Understanding

Find the product. Use base-ten blocks or draw intersecting lines on your own paper if you wish.

1 17 × 3 = ■

2 16 × 8 = ■

Chapter 12 Lesson 5
REVIEW MODEL
Using Rectangles to Represent Arrays

You can think about arrays of intersecting lines to help you multiply by a two-digit number.

Example Find 18 × 7.

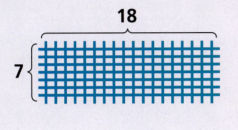

You can think about an array with 7 horizontal lines and 18 vertical lines.

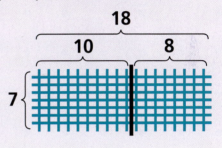

You can think about the best way to separate the array, so you have simpler problems to solve.

Then you can use a shortcut to record your solution and find the product.

Step ❶ You can represent the array using a separated rectangle instead of drawing all the lines.

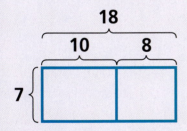

Step ❷ You can record the number of intersections in each section.

7 × 10 = 70
7 × 8 = 56

Step ❸ You can find the number of intersections in the complete array by adding the amounts in each section.

70 + 56 = 126
So, 18 × 7 = 126.

✓ Check for Understanding

Draw a diagram to help you find the product.

❶ 19 × 6 = ■

❷ 27 × 5 = ■

Chapter 12
Lesson 6
EXPLORE
Separating Arrays

This array shows 15 × 17. Use straws to separate the array and find the number of dots in the array.

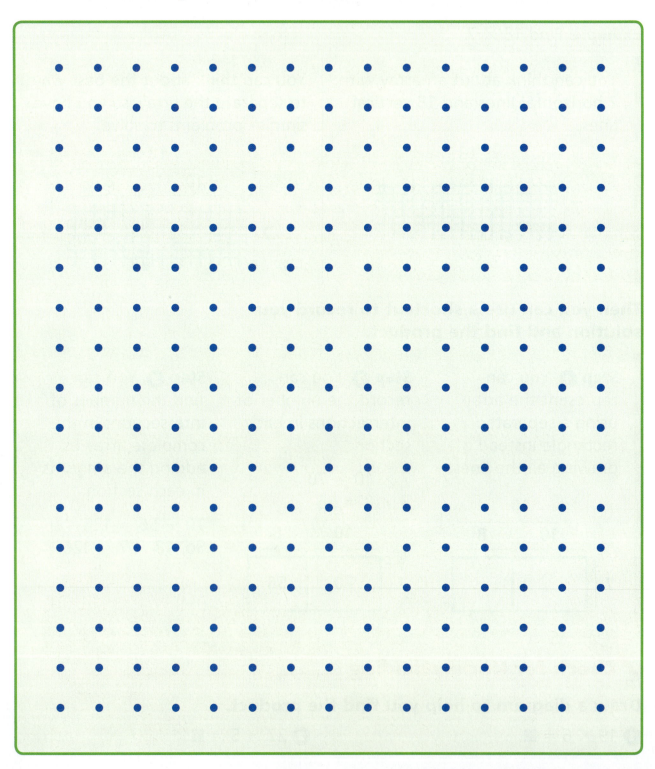

Chapter 12
Lesson 8
EXPLORE
Division Situation

A crate of 96 juice boxes was delivered to the cafeteria. It contained apple, orange, grape, pineapple, cranberry, and tomato juice. If there was an equal number of each type of juice, how many grape juice boxes were delivered?

Solve the problem and write a number sentence to match. Then write the related number sentences. Use tiles, grid paper, or draw a picture to help you. Be prepared to explain your solution.

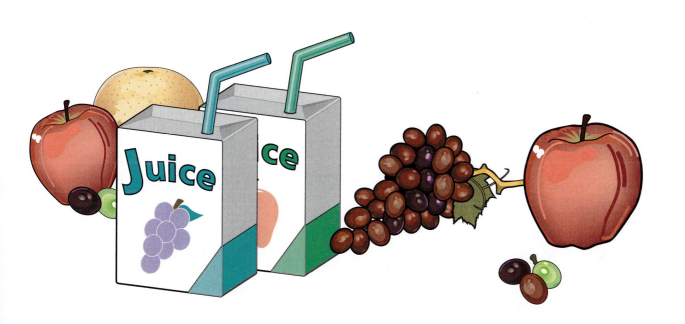

Chapter 12 Lesson 9
REVIEW MODEL
Problem Solving Strategy
Work Backward

Hector wants his new friend to guess his age. He tells his friend that he multiplied his age by 5, and then added 8 to the result. His final answer was 83. What is Hector's age?

___ × 5 + 8 = 83

Strategy: Work Backward

 Read to Understand

What do you know from reading the problem?

When Hector multiplies his age by 5 and adds 8, he gets 83.

 Plan

How can you solve this problem?

You can use the strategy *work backward*.

 Solve

How can you use this strategy?

You can start with Hector's final answer of 83. Next, subtract the 8 he added:

83 − 8 = 75.

Then solve 5 × ■ = 75 to find his age.

5 × 15 = 75   So, Hector is 15 years old.

 Check

Look back at the problem. Did you answer the question that was asked? Does the answer make sense?

Problem Solving Practice

Problem Solving Strategies
- ✔ Act It Out
- ✔ Draw a Picture
- ✔ Guess and Check
- ✔ Look for a Pattern
- ✔ Make a Graph
- ✔ Make a Model
- ✔ Make an Organized List
- ✔ Make a Table
- ✔ Solve a Simpler Problem
- ✔ Use Logical Reasoning
- ✔ **Work Backward**
- ✔ Write a Number Sentence

Use the strategy *work backward*.

① Omar paid $14 for 4 sandwiches and 2 drinks. What was the cost of each drink?

> Sandwiches $2.50
> Salads $1.75
> Drinks

② Yvette sold a fourth of her doll collection at the garage sale. She sold 20 dolls. How many dolls did Yvette have before the garage sale?

Mixed Strategy Practice

Use any strategy to solve. Explain.

③ Jorge has 36 strawberries divided equally into 4 containers. How many strawberries are in 2 containers?

④ Leticia used nickels and dimes to pay for a toy that cost $1.00. If she used 2 more nickels than dimes, how many dimes did she use?

⑤ If a butterfly flaps its wings about 400 times a minute, about how many times will it flap its wings in 3 minutes? How many times in 5 minutes?

⑥ The Ferris wheel at the amusement park has 20 cars. Each car can hold up to 8 people. What is the largest number of people that can ride the Ferris wheel at one time?

⑦ The box shown at right is made up of green and blue centimeter cubes. All six sides look the same. How many blue centimeter cubes were used to make the box?

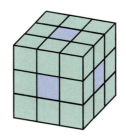

Chapter 12

Chapter 12 Vocabulary

Choose the best vocabulary term from Word List A for each sentence.

Word List A

array
column
diagonal
dividend
divisor
factor
horizontal
intersection
output
pattern
product
quotient
rule
shorthand notation
sum
vertical

1. The direction that goes from top to bottom is __?__.

2. A(n) __?__ is an operation on a number.

3. A(n) __?__ for an array is the line from an upper corner toward the opposite lower corner.

4. The __?__ is where two lines cross.

5. The number that results from dividing is the __?__.

6. A rectangular arrangement that shows objects in rows and columns is called a(n) __?__.

7. A number that is multiplied by another number is called a(n) __?__.

8. The __?__ is the result of multiplication.

Complete each analogy using the best term from Word List B.

Word List B

array
input
row
rule

9. Toast is to bread as output is to __?__.

10. Vertical is to column as horizontal is to __?__.

Talk Math

Discuss with a partner what you have learned about multiplication. Use the vocabulary terms *factor*, *product*, and *sum*.

11. How can you separate a number to multiply?

12. How can you use base-ten blocks to multiply two numbers?

13. How can you use shorthand notation instead of an array to multiply?

Analysis Chart

14. Create an analysis chart for the terms *dividend, divisor, factor, product, quotient,* and *sum.* Use what you know and what you have learned about operations with whole numbers.

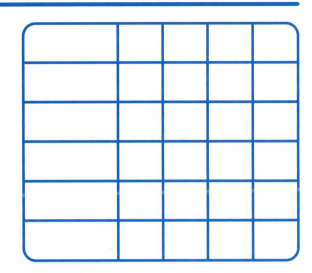

Word Web

15. Create a word web using the term *rule*. Include similar words you know and what you have learned about math rules.

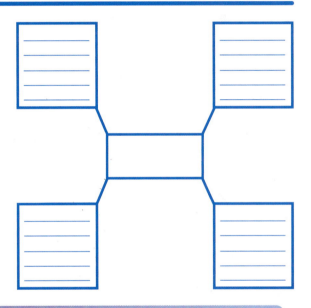

What's in a Word?

PATTERN A *pattern* can be many things. It can be a design that repeats. You see patterns like that on wallpaper, gift wrap, and clothing. A *pattern* can also be a model or a guide for making something. A tailor uses that type of *pattern,* because it shows the shape and size of the pieces of cloth to cut out. A clothing *pattern* is a sort of "rule" to follow for making a piece of clothing.

In math, *patterns* also follow a rule. A math *pattern* can be an ordered set of numbers or figures.

Technology
Multimedia Math Glossary
www.harcourtschool.com/thinkmath

GAME

Multiplication Challenge

Game Purpose
To multiply with factors to 18

Materials
- 3 number cubes
- Activity Master 164: Multiplication Challenge
- Activity Master 165: Multiplication Table: Factors to 18 or a calculator
- Base-ten blocks or counters (optional)

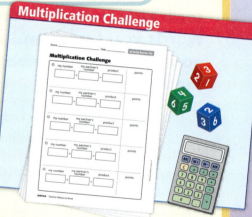

How To Play The Game

1. This is a game for 2 players. Each player will need a copy of Activity Master 164 and Activity Master 165. You may use a calculator instead of Activity Master 165.

2. Take turns tossing the 3 number cubes.
   - Find the sum of the 3 numbers you tossed.
   - Record that number on Activity Master 164.
   - Then record the other player's number.

3. By yourself, multiply the numbers. Record the product.

4. Together, decide what the correct product is. Each player with a correct product gets one point.

5. Repeat steps 2 to 4 until you have completed 5 problems.

6. Check your answers using the multiplication table or a calculator. You might have to adjust your score. The player who has more correct answers wins.

Factor Tic-Tac-Toe

Game Purpose
To practice strategies for multiplying with two-digit numbers

Materials
- Activity Master 166, 167, 168, or 169: Factor Tic-Tac-Toe gameboards
- Two-color counters
- Calculator

How To Play The Game

1. This is a game for 2 players. Choose your *Factor Tic-Tac-Toe* gameboard. Decide who will use each color of counter.

2. Player 1 chooses a number on the gameboard. Player 2 chooses a different number.

3. Each player finds the product of the two numbers. Record your work on a separate sheet of paper.

4. Use the calculator to check your answers. If your answer is correct, place a counter over the number you chose to multiply. Both players, only one player, or neither players may be able to place a counter.

5. Repeat steps 2 to 4. Take turns being the first player to choose a number. The first player to get 3 counters in a row, column, or diagonal wins.

6. Play again. Use the same gameboard, or choose a different one. Play as many games as time allows.

CHALLENGE

Use each number in the box one time to complete the problems. Then solve each problem. When you are finished, each problem will make sense. Each answer will be a whole number. *Hint:* Decide which problems can be solved by dividing. Do those first.

| 24 | 33 | 6 | 8 | 7 | 7 |
| 9 | 91 | 85 | 64 | 168 |

1. A train has ■ cars. Each car carries ■ passengers. There are between 250 and 300 passengers. How many passengers does the train carry in all?

2. A juice carton holds ■ ounces. Each serving of juice is ■ ounces. The number of servings is the same as the number of ounces in each serving. How many servings can be made from a full carton?

3. Tammy's Toy Store sells bags of marbles. There are ■ marbles in each bag. There are ■ bags in the store. There are between 550 and 650 marbles in the store. How many marbles are in the store?

4. Roberto rode his bicycle ■ miles last week. He rode every day except Thursday. He rode the same number of miles each day. The number of miles he rode each day was greater than 20. How many miles did he ride each day?

5. There are ■ students in the school band. They march in rows of ■ students. How many rows of students are there?

6. The library received ■ cartons of books. There are ■ books in each carton. If there are fewer than 150 books, how many books are in the cartons?

Chapter 13
Time, Temperature, Weight, and Capacity

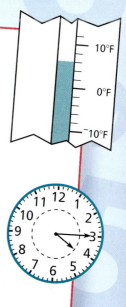

Dear Student,

If someone says it's 80°F outside, do you need a jacket? Or is it hot outside?

Some things can be measured, but not with a ruler. Once you have some experience measuring the temperature, you can tell someone what the temperature is. You will know whether it's cold, hot, or just right. In this chapter, you will measure different things, using tools such as clocks, thermometers, scales, and measuring cups.

You will use several skills to make sense of all the things you measure, including:

- understanding what you are measuring,
- choosing the right kind of tool to measure,
- using and reading the tool,
- knowing what measurement numbers mean, and
- drawing conclusions from measurements.

Get ready to measure!

Mathematically yours,
The authors of **Think Math!**

THE WORLD ALMANAC FOR KIDS

Winter Pleasures: Cold and Hot

In snowy places in the winter, many children make snowmen. It can snow a lot in Boston, Massachusetts. A typical temperature in Boston in February is 32°F. Fairbanks, Alaska, is even snowier and colder. A typical temperature in Fairbanks in February is ⁻4°F.

FACT·ACTIVITY 1

Use the temperatures to answer the questions.

1. What temperature does the thermometer for Boston show? What temperature does the Fairbanks thermometer show?

2. Is the Boston temperature on the thermometer higher or lower than Boston's typical February temperature? How much higher or lower?

3. Is the Fairbanks temperature on the thermometer higher or lower than it's typical February temperature? How much higher or lower?

Use the times to answer the questions.

4. It is morning in Fairbanks. Write the time. What time will it be in 5 hours?

5. It is daytime in Boston. Suppose the temperature in Boston increases 3°F an hour for the next 5 hours. What will the temperature be in 5 hours? What will the time be?

6. A snowman will begin to melt when the temperature is 32°F or above. Will the snowman in Boston begin to melt in the next 5 hours? Explain.

After playing in the snow, children and adults might enjoy hot soup or hot cocoa. Sammy's mother made 2 gallons of soup for a hungry crowd. However, the soup pot is too big to store in the refrigerator. Help Sammy's mom figure out how she can refrigerate the soup.

FACT·ACTIVITY 2

Write yes or no for each choice. If no, tell how many more and what containers are needed.

1. Can the soup be stored in 3 half-gallon jugs?
2. Can the soup be stored in 8 one-quart containers?
3. Can the soup be stored in 4 one-quart containers and 4 one-pint jars?

CHAPTER PROJECT

Find out how long it takes for a cup of cold water and a cup of tap water to warm to room temperature.

- Fill a 4-ounce plastic cup with water from the tap. Make it colder by refrigerating it for 4 hours.
- When the 4 hours are up, fill a second cup with 4 ounces of tap water.
- Use a thermometer to measure the temperature of each cup of water.
- Measure the temperature every 30 minutes until each cup reaches the air temperature in your room. Record the temperatures and time in a table.
- Make a graph to show your results.
- Analyze the results. Compare the amount of time it took for each cup of water to reach room temperature.

ALMANAC Fact

To create a snow sculpture, experts usually begin with a single block of snow about 6 to 15 feet on each side. The winner of the 2006 International Snow Sculpture Competition was a puppy looking in a mirror.

Chapter 13
Lesson 3
EXPLORE
How Much Time Is Left?

> Snaily needs to be home at **5:45** P.M. She is worried that she will be late, so she looks at her watch several times.

1 For each picture of Snaily's watch, write the time and the number of minutes left until **5:45**.

A 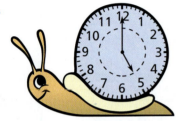 5:00, 45 minutes left

B

C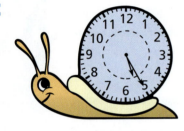

D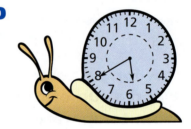

2 How many minutes passed without Snaily looking at her watch . . .

. . . between times A and B?

. . . between times B and C?

. . . between times C and D?

3 Snaily can travel **1 inch** in **5 minutes**.

A How far can she go in **10 minutes**?

B How far can she go in a half hour?

C If Snaily was **10 inches** away from home at **4:50** P.M., could she make it home on time? Explain.

Chapter 13 Lesson 3
REVIEW MODEL
Time, Distance, and Speed

You can draw a picture to help you find how far you can travel or how much time it takes.

It takes Matt 10 minutes to walk his dog 1 block. If he continues walking at the same speed, how far can Matt walk his dog in 20 minutes?

Think:

20 minutes − 10 minutes = 10 minutes

In 10 minutes, Matt can walk his dog 1 block.

In the remaining 10 minutes, he can walk his dog another block.

So, Matt can walk his dog 2 blocks in 20 minutes.

You may draw a picture like this to help you.

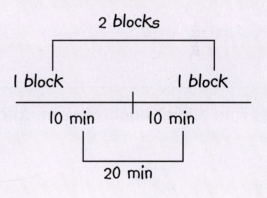

✓ Check for Understanding

You may draw a picture to help answer the questions.

1 How long would it take Matt to walk his dog 3 blocks?

2 The park is 2 blocks away from Matt's house. How long will it take Matt to walk his dog to the park and back?

3 How many blocks can Matt and his dog walk in 1 hour?

Chapter 13
Lesson 5 — EXPLORE
Estimating Weight

How many pounds of books do you have in your classroom?

1 How many books do you have in your desk?

2 About how many books are in your classroom?

3 Find a **light** book. How much does it weigh?

4 Find a **heavy** book. How much does it weigh?

5 Find an **average** book—one that is not very light and not very heavy. How much does it weigh?

6 About how much do you think all the books in your classroom weigh?

Be ready to explain how you made your estimates.

Chapter 13 Lesson 6
REVIEW MODEL
Comparing Capacities

You can use a diagram to help you write the same capacity in different ways.

This diagram can help you find how many cups, pints, and quarts are in 1 gallon.

> Different letters in the diagram show different units of capacity. The size of each letter is important.

The largest letter, G, is 1 gallon. The gallon is the largest unit of capacity in the diagram. It helps you see that one G (gallon) holds more than one Q (quart).

How many quarts does 1 gallon hold?
Count the quarts.
There are 4 quarts (Q) in 1 gallon (G).

So, 4 quarts = 1 gallon.

G = gallon
Q = quart
P = pint
C = cup

You can use the diagram to help you write the same capacity in different ways.

Follow these steps to show the number of cups in 1 quart:

Step 1 Find the Q (quart).

Step 2 Count the number of cups (C) inside the Q (quart).

Step 3 Write the equal capacities.

4 cups = 1 quart

✓ Check for Understanding

Use the diagram to help you write the same capacity in different ways.

1. How many pints are in 1 gallon?

2. How many cups are in 1 gallon?

3. How many pints are in 2 quarts?

4. Jill brought 4 quarts of water to the game. Brandon brought 6 pints. Who brought more water? How do you know?

Chapter 13 **205**

Chapter 13 Lesson 9
REVIEW MODEL
Problem Solving Strategy
Act It Out

The Art Club had a party. They served 3 pounds of cheese. The guests ate 29 ounces of the cheese. How much cheese was left?

Strategy: Act It Out

Read to Understand

What do you know from reading the problem?

There were 3 pounds of cheese. The guests ate 29 ounces. You need to find how much cheese was left.

Plan

How can you solve this problem?

You can act it out by using objects, such as counters, to represent the number of ounces.

Solve

How can you act it out?

To find how many ounces of cheese are left, you need to know how many ounces are in 3 pounds of cheese.

Since there are 16 ounces in 1 pound, you can count out 3 groups of 16 counters. 3 × 16 = 48 counters, so 3 pounds = 48 ounces.

Then subtract 29 counters for the number of ounces that were eaten.

48 − 29 = 19, so there were 19 ounces of cheese left.

Check

Look back at the problem. Did you answer the question that was asked? Does the answer make sense?

Problem Solving Practice

Use the strategy *act it out*.

1 The Art Club made 16 cups of berry punch. The guests drank 96 fl oz of punch. How many ounces of punch were left?

2 At 6:00 P.M. when the snowstorm began, the temperature was 4°F. At 9:00 P.M., it was 12°F colder. What was the temperature at 9:00 P.M.?

Problem Solving Strategies

✓ **Act It Out**
✓ Draw a Picture
✓ Guess and Check
✓ Look for a Pattern
✓ Make a Graph
✓ Make a Model
✓ Make an Organized List
✓ Make a Table
✓ Solve a Simpler Problem
✓ Use Logical Reasoning
✓ Work Backward
✓ Write a Number Sentence

Mixed Strategy Practice

Use any strategy to solve. Explain.

3 The Drama Club put on a play for the school. It started at 2:15 P.M. and lasted 1 hour 45 minutes. What time did the play finish?

4 It takes Jenna 6 minutes to walk 3 blocks. Her house is 9 blocks from her school. If she leaves school at 3:00 P.M., what time will Jenna get home?

For 5 and 6, use the table.

5 Which day had the warmest afternoon temperature?

6 How would you describe the trend in the temperatures from Monday to Thursday?

Day	Temperature	
	Morning	Afternoon
Monday	62°F	78°F
Tuesday	63°F	79°F
Wednesday	68°F	84°F
Thursday	70°F	86°F

Chapter 13   207

Chapter 13 Vocabulary

Choose the best vocabulary term from Word List A for each sentence.

❶ A customary unit for measuring something that weighs less than a pound is a(n) __?__.

❷ Any number that is less than zero is a(n) __?__.

❸ __?__ is a measure of how hot or cold it is.

❹ The amount of time that passes from the start of an activity to the end of that activity is __?__.

❺ The amount that a container can hold is called its __?__.

❻ A(n) __?__ is a customary unit of capacity that is the same as 2 pints.

Word List A

capacity
cup
decrease
elapsed time
Fahrenheit
fluid ounce
gallon
increase
negative number
negative sign
ounce
pint
pound
quart
temperature
ton

Complete each analogy using the best term from Word List B.

❼ Length is to ruler as temperature is to __?__.

❽ Clock is to time as __?__ is to weight.

Word List B

balance
estimate
thermometer

💬 Talk Math

Discuss with a partner what you have learned about measurement. Use the vocabulary terms *decrease, increase, earlier,* and *later.*

❾ How can you find a change in temperature on a thermometer?

❿ How can you find elapsed time on an analog clock?

Degrees of Meaning Grid

11. Create a degrees of meaning grid for *capacity*, *temperature*, *time*, and *weight*. Use what you know and what you have learned about measurement.

General	Less General	Specific	More Specific

Analysis Chart

12. Create an analysis chart using the words *cup*, *pint*, *quart*, and *gallon*.

What's in a Word?

POUND A car might be towed to the *pound* for parking in the wrong place. A *pound* is a place where stray animals are taken. In England, a *pound* is a unit of money. You can *pound* a nail with a hammer. You can *pound* your fist on a door. You can eat *pound* cake, which got its name because the original recipe called for one *pound* of each ingredient. In math, a *pound* is a customary unit of weight equal to 16 ounces.

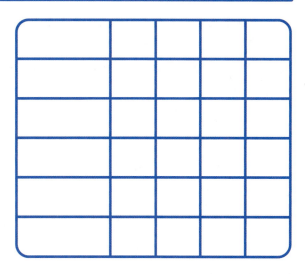

GO ONLINE Technology
Multimedia Math Glossary
www.harcourtschool.com/thinkmath

Chapter 13 **209**

GAME

The Freezing Game

> **Game Purpose**
> To practice reading temperatures on a thermometer
>
> **Materials**
> - Activity Master 170: The Freezing Game
> - 4 counters
> - 2 number cubes

How To Play The Game

1 This is a game for 2 players. The goal is to land on 32°F as many times as you can. Use the *Freezing Game* Activity Master. Choose your color counter. Put 1 counter at the freezing point for water (32°F) to mark the temperature. Put 1 counter in your 0 box to count the times you land at freezing. Toss the number cubes to see who goes first. After that, take turns.

2 Toss both number cubes.
  - Find the sum or difference of the two numbers.
  - Move your temperature counter that many spaces up or down on the thermometer.
  - You must move your counter.

Example: You get these numbers. You may move up or down 7 degrees. You may move up or down 3 degrees.

But if you get these numbers, you must move up or down 6 degrees. You may not use the difference of 0 to stay where you are.

3 Each time you land on 32°F at the end of your turn, move your counting counter to the next box. The first player to land on the freezing point 4 times is the winner.

Time Concentration

Game Purpose
To practice telling time

Materials
- Activity Master 171: Time Concentration Cards (Deck 1)
- Scissors

How To Play The Game

1 This is a game for 2 players. The game is played like any other *Concentration* game. The object is to match times shown on an analog clock to times written as they would appear on a digital clock.

2 Cut out the *Time Concentration Cards* from Deck 1. Mix up the cards. Place them face down to form a 4-by-6 rectangular array. Decide who will go first. After that, take turns.

3 Turn over two cards.
- If both cards show the same time (one analog and one digital), it's a match. Take the cards. You get another turn.

- If the cards do not match, turn them face down again. Your turn is over.

4 When all the matching pairs have been found, count your cards. The player with more cards is the winner. (There could be a tie.) If there is time, mix up all the cards again, and play another game.

CHALLENGE

Ms. Clark gave each student in her class a paper bag with marbles hidden inside. Each bag has a different number of marbles. The students made up measurement riddles about the number of marbles in their bags.

How many marbles does each student have? Solve each riddle to find out.

1. Drake has as many red marbles as there are quarts in a half gallon. He has as many blue marbles as there are cups in 2 pints. How many marbles does Drake have?

2. Joelle has as many green marbles as there are ounces in a half pound. She has as many red marbles as there are tons in 4,000 pounds. How many marbles does Joelle have?

3. Calvin has as many blue marbles as there are pints in $2\frac{1}{2}$ quarts. He has as many orange marbles as there are fluid ounces in $\frac{1}{2}$ cup. How many marbles does Calvin have?

4. Aiko has as many clear marbles as there are quarts in 16 cups. She has as many yellow marbles as there are gallons in 8 quarts. How many marbles does Aiko have?

5. Scott has as many green marbles as there are cups in $1\frac{1}{2}$ pints. He has as many orange marbles as there are quarts in $1\frac{1}{2}$ gallons. How many marbles does Scott have?

6. Donette has as many red marbles as there are pints in 2 gallons. She has as many blue marbles as there are cups in 2 quarts. How many marbles does Donette have?

7. Javon has as many green marbles as there are pounds in 48 ounces. He has as many blue marbles as there are ounces in $1\frac{1}{2}$ pounds. He has as many clear marbles as there are tons in 6,000 pounds. How many marbles does Javon have?

Chapter 14
Addition and Subtraction in Depth

Dear Student,

When we need to solve difficult math problems, we sometimes use related problems that are simpler to solve.

> For example, to solve 70,000 + 60,000 = ■, we might use 7 + 6 = ■. How could 7 + 6 = ■ help with 70,000 + 60,000 = ■? Which other problems that involve large numbers could you solve using 7 + 6 = ■?

The idea of place value makes it possible to use 7 + 6 = ■ to complete 70,000 + 60,000 = ■. Can you explain how place value can help you solve problems with larger numbers?

In this chapter, you will learn more ways to solve challenging addition and subtraction problems by using simpler problems. The idea of place value will be at the center of these ways.

So go ahead! Use the tools you have to make "difficult" problems simpler to solve.

Mathematically yours,
The authors of **Think Math!**

THE WORLD ALMANAC FOR KIDS

A Visit to New York City

New York City has the largest population of any city in the United States. Five parts, called boroughs, make up New York City. Two of the boroughs are islands. If you visit the city, you can get around by bus, taxi, subway, car, and even by boat!

FACT·ACTIVITY 1

Major Transportation Systems in New York City				
Type of Transportation	Buses	Licensed Taxis	Subway Cars	Ferries
Total Number	4,489	12,778	?	5

For 1–4, use the table.

1. What is the place value of the digit 1 in the total number of licensed taxis?

2. Describe the total number of licensed taxis in words.

3. The buses in New York City run 44,550 trips on a daily basis. Write the number of buses and the number of trips in expanded form.

4. The total number of running subway cars is missing from the table. Its thousands digit is 3 times its hundreds digit. Which of the following could be the number of running subway cars: 6,030; 6,200; 620; 2,600?

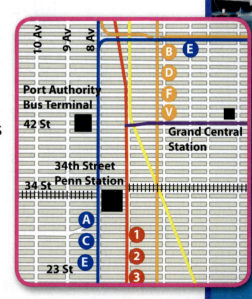

214 Chapter 14

The subway system has *underground* tracks below street level, street level tracks, *embankment* tracks built up on earth or stone, and elevated tracks built on high platforms above street level.

FACT·ACTIVITY 2

Just for fun, some people try to ride the entire New York subway system in the shortest time possible. In August, 2006, two college students finished the ride in 24 hours 2 seconds!

For 1–3, use the table.

New York City Subway Facts

Type of Stations	Number of Stations
Underground	277
Elevated	?
Embankment	29
Street level	9
TOTAL	**468**

1. How many underground, embankment, and street level stations are there in all?
2. Find the number of elevated stations.
3. New York subway stations are also famous for their interesting art and designs. Mike wants to see the art in each subway station. He rides to 152 stations on the first day and 137 more stations on the second day. How many more stations are left for Mike to see?

CHAPTER PROJECT

Central Park is a must-see attraction when you visit New York. Create a brochure that includes the area of different parts of the park and the distances to the park from other major attractions. Also include:

- a map of the park
- directions to the park
- interesting facts about the park, such as total number of water fountains, benches, ponds, or species of birds.

Write three 2- and 3-digit addition and/or subtraction sentences based on your facts.

ALMANAC Fact

There are about 6,375 miles of streets in New York City. There is no street in Manhattan called Main Street.

Chapter 14
Lesson 2
EXPLORE
Exploring Multi-Digit Addition

1 Make up a situation to go with the problem:

$$163 + 24$$

Write each number in expanded form. Find a way to use the expanded form to add the numbers. If it helps, use base-ten blocks or draw a diagram. Be prepared to explain your solution.

---

2 Ruby added a number to 384 and got a sum with a 6 in the tens place.

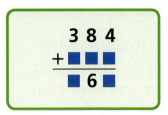

Could her sum be correct? Explain.

Chapter 14 Lesson 2
REVIEW MODEL
Using Expanded Form to Add

You can break up numbers to add. Breaking up numbers can help you regroup the place values to find the sum.

Example Find the sum.   365
                          + 289

Step 1 Break up both addends by writing them in expanded form.

$365 = 300 + 60 + 5$

$289 = 200 + 80 + 9$

Step 2 Add each place value.

$365 = 300 + 60 + 5$
$+ 289 = 200 + 80 + 9$
$ = 500 + 140 + 14$

Step 3 Regroup if there is 10 or more in the ones place.

Rewrite 14 using expanded form: $14 = 10 + 4$

Add the 10 to the tens place. Put the remaining 4 ones in the ones place.

$365 = 300 + 60 + 5$
$+ 289 = 200 + 80 + 9$
$ = 500 + 150 + 4$

Step 4 Regroup if there is 100 or more in the tens place.

Rewrite 150 using expanded form: $150 = 100 + 50$

Add the 100 to the hundreds place. Put the remaining 50 in the tens place.

$365 = 300 + 60 + 5$
$+ 289 = 200 + 80 + 9$
$ = 600 + 50 + 4$

Step 5 Add the sums in each place value to find the total sum.

  365
+ 289

  654

✓ Check for Understanding

Find the sum.

1 78 + 22

2 403 + 168

3 655 + 267

4 248 + 109

5 66 + 36

6 782 + 235

Chapter 14
Lesson 3
EXPLORE
Exploring Multi-Digit Subtraction

Four students broke up **348** in different ways: 〰 **348**

Cliff's way:
300 + 40 + 8

Sean's way:
200 + 130 + 18

Paula's way:
200 + 140 + 8

Lana's way:
300 + 30 + 18

For Problems 1 and 2, use base-ten blocks or draw a picture to explain your thinking.

1 Which way is most useful for subtracting **172** from **348**?

2 Which way is most useful for subtracting **89** from **348**?

How would you break up 232 if you were subtracting . . . 〰 **232**

3 11?   **4** 96?   **5** 113?

6 Make up a situation to go with one of the subtraction problems on this page. Write a number sentence to go with your story.

218 Chapter 14

Chapter 14
Lesson 3

REVIEW MODEL
Using Expanded Form to Subtract

You can break up numbers to subtract. Then you can regroup, if necessary, to find the difference.

Example Find the difference.   755
                                 − 167

Step 1 Write each number in expanded form.   755 = 700 + 50 + 5
                                                 167 = 100 + 60 + 7

Step 2 Subtract the ones.

Think: 5 < 7

So, take 10 from the tens place of the larger number and add it to the ones place.

The tens place of the larger number changes from 50 to 40, and the 5 ones become 15 ones.

$$755 = 700 + 40 + 15$$
$$- 167 = 100 + 60 + 7$$
$$8$$

Step 3 Subtract the tens.

Think: 40 < 60

So, take 100 from the hundreds place of the larger number and add it to the tens place.

The hundreds place of the larger number changes from 700 to 600, and the tens place changes from 40 to 140.

$$755 = 600 + 140 + 15$$
$$- 167 = 100 + 60 + 7$$
$$80 + 8$$

Step 4 Subtract the hundreds. Then add the numbers in each place value to find the difference.

$$755 = 600 + 140 + 15$$
$$- 167 = 100 + 60 + 7$$
$$500 + 80 + 8 = 588$$

✓ Check for Understanding

Find the difference.

1. 426 − 118
2. 356 − 124
3. 176 − 29
4. 142 − 65
5. 521 − 433
6. 257 − 183

Chapter 14
Lesson 4

EXPLORE
Exploring Addition and Subtraction

How would you break up these numbers to add and subtract?

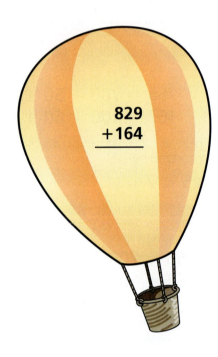

829
+164

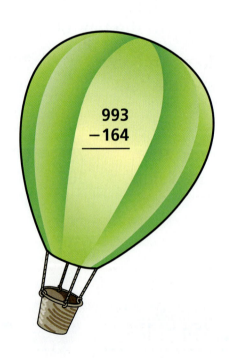

993
−164

Chapter 14 Lesson 6
EXPLORE
Exploring Situations

David wants to know if there are more science books or biography books in the library.

In the science section, there are 48 books on the first shelf, 65 books on the second shelf, and 77 books on the third shelf.

In the biography section, there are 105 books on the first shelf, 52 books on the second shelf, and 68 books on the third shelf.

1 Which section has more books?

2 How many more books are in that section than in the other?

3 Be prepared to discuss how you solved the problem.

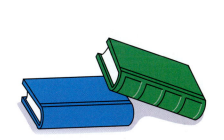

Chapter 14 Lesson 7

REVIEW MODEL
Problem Solving Strategy
Solve a Simpler Problem

Shanna owns a garden shop. She spent $53 for each small clay pot and sold each one for $80. She spent $112 for each medium clay pot and sold each one for $145. She spent $137 for each large clay pot and sold each one for $170. If Shanna sold one clay pot of each size on one day, how much money did she make that day?

$80 $145 $170

Strategy: Solve a Simpler Problem

Read to Understand

What do you know from reading the problem?

The amount Shanna paid for each size of clay pot and the sale price of each size of pot.

Plan

How can you solve this problem?

You can use the strategy *solve a simpler problem.*

Solve

How can you use this strategy to solve the problem?

Find the amount she made on each size of pot by finding the difference between the amount she paid and the sale price for each pot.

Small clay pot: $80 − $53 = $27

Medium clay pot: $145 − $112 = $33

Large clay pot: $170 − $137 = $33

Then find the total amount she made.

$27 + $33 + $33 = $93   So, Shanna made $93.

Check

Look back at the problem. Did you answer the question that was asked? Does the answer make sense?

Problem Solving Practice

Use the strategy *solve a simpler problem*.

1. Paul is in charge of counting students as they enter and leave the school carnival. He counted 455 students entering before 1:00 P.M. He counted 126 students leaving at 2:00 P.M. At 3:00 P.M., 56 more students arrived and 14 students departed. At 4:00 P.M., 111 students left and 44 students arrived. How many students were at the carnival at 4:00 P.M.?

2. Carrie mows 14 lawns each week. How many lawns does she mow in 9 weeks?

Mixed Strategy Practice

> **Problem Solving Strategies**
> ✓ Act It Out
> ✓ Draw a Picture
> ✓ Guess and Check
> ✓ Look for a Pattern
> ✓ Make a Graph
> ✓ Make a Model
> ✓ Make an Organized List
> ✓ Make a Table
> ✓ **Solve a Simpler Problem**
> ✓ Use Logical Reasoning
> ✓ Work Backward
> ✓ Write a Number Sentence

Use any strategy to solve. Explain.

3. What is the missing output?

Input	5	7	3	9	4
Output	14	16	12	18	■

4. What is the mystery number? The number is greater than 40 but less than 50. The number is odd. When you add the two digits together, you get a sum of 13.

5. Rod has 42 stickers to share evenly with his 6 friends. How many stickers will each person get if Rod shares his stickers with his friends and gives himself the same number of stickers?

6. At 5:00 P.M. when the baseball game started, the temperature was 63°F. By 7:00 P.M., it was 8 degrees cooler. By 9:00 P.M., the temperature had dropped another 16 degrees. What was the temperature at 9:00 P.M.?

7. Brooke arranged her picture frames on her bedroom wall in an array with 8 rows and 3 columns. What other ways could she arrange the same number of pictures if she wants to make an array, but does not want to have all the pictures in one row or one column?

8. Mr. Hood plans to work in his garden 12 hours during the next 3 days. He wants to work for half of the planned time on Friday, half of the time that is left on Saturday, and the remaining time on Sunday. How many hours does Mr. Hood plan to work in his garden on Sunday?

Chapter 14 Vocabulary

Choose the best vocabulary term from Word List A for each sentence.

Word List A
- addend
- compatible numbers
- difference
- estimate
- expanded form
- multi-digit number
- regroup
- round
- standard algorithm
- sum

1. A number close to an exact amount is a(n) __?__.

2. A(n) __?__ has more than one digit.

3. When you write a number to its nearest ten or hundred you __?__.

4. A number that is added to another in an addition problem is called a(n) __?__.

5. A way to write a number to show the value of each digit is called __?__.

6. To __?__ means to exchange equal amounts when working with a number.

7. The answer in a subtraction problem is called the __?__.

Complete each analogy using the best term from Word List B.

Word List B
- difference
- estimate
- sum

8. Compatible numbers are to estimate as addends are to __?__.

9. Addition is to subtraction as sum is to __?__.

Talk Math

Discuss with a partner what you have learned about addition and subtraction. Use the vocabulary terms *expanded form, regroup, standard algorithm,* and *sum.*

10. How can you add two multi-digit numbers?

11. How can you use sums to 10 to add a list of two-digit numbers?

Word Definition Map

12 Create a word definition map for the term *estimate*.

  A What is it?

  B What is it like?

  C What are some examples?

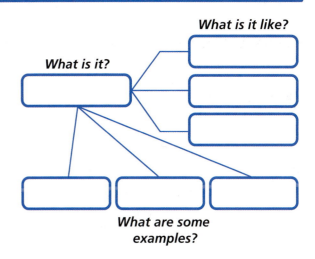

Word Web

13 Create a word web using the term *sum*.

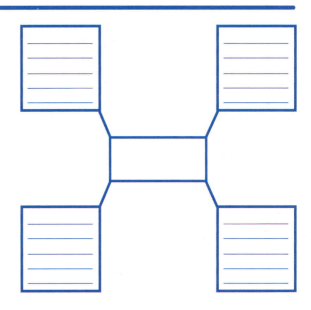

What's in a Word?

ESTIMATE This word is spelled one way, but it can be pronounced two ways. When it is a noun (a naming word), the *A* in the last syllable has a short sound. It sounds like the *I* in the middle syllable. "The *estimate* of the sum is 600." The verb (action word) has a long *A* sound in the last syllable. It sounds like the word for friend, *mate*. "*Estimate* the sum of 362 and 231."

GO ONLINE Technology
Multimedia Math Glossary
www.harcourtschool.com/thinkmath

Chapter 14 **225**

GAME

Place Value Game

Game Purpose
To practice reading and comparing place values of six-digit numbers

Materials
- Activity Masters 177–184: Attribute Cards
- Scissors

How To Play The Game

1 This is a game for 3, 4, or 5 players. The goal is to match 5 six-digit numbers that you choose to attributes on the Attribute Cards. The blue cards are the easiest. The green cards are more challenging. As a group, decide which color cards to use. Or you may use all of the cards.

2 Cut out the Attribute Cards. Mix them up. Place them face down in a stack.

3 Write 5 six-digit numbers on a sheet of paper. Write neatly and large enough for the others to see.

4 Take turns. Turn over one attribute card at a time. Read it aloud. Cross out any of your numbers that match the attribute. You may help one another decide which numbers to cross out.

5 The first player to cross out all 5 numbers wins. If there is time, play another game.

Addition Scramble

Game Purpose
To practice addition with multi-digit numbers

Materials
- Activity Master 187: Addition Scramble Game Page
- Number cards (1–9, four sets)

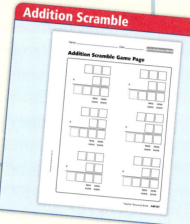

How To Play The Game

1. This is a game for 2 players. The goal is to score fewer points. Use the same game page.

2. Mix up the number cards. Place them face down in a stack.
   - One player takes 3 cards. He or she puts the cards in any order to make a three-digit number.
   - The other player takes 2 cards. He or she makes a two-digit number.
   - Both players record their numbers on the Addition Scramble Game Page.

3. Find the sum of the numbers.

4. The first player's score is the digit in the tens place. The second player's score is the digit in the ones place.

5. Trade roles. Keep a running total of your scores. Play until one player reaches 50 points. The other player wins!

CHALLENGE

Solve these ten number puzzles. Work backward from the starting number to find each number. Each puzzle will involve addition, subtraction, or comparisons. Some puzzles might use two of those or all three. The first puzzle is set up for you.

1. What number is 8 more than 20 more than 5 more than 60?
2. What number is 25 more than 9 more than 32 more than 14 more than 18?
3. What number is 3 less than 2 more than 17 more than 11?
4. What number is 5 less than 4 less than 6 less than 59?
5. What number is 12 more than 1 more than 3 less than 25 more than 6?
6. What number is 10 less than 17 more than 6 more than 2 less than 21?
7. What number is 4 more than 15 less than 19 more than 8 less than 52?
8. What number is 9 less than 23 less than 14 more than 16 more than 73?
9. What number is 27 more than 8 less than 16 less than 10 less than 35?
10. What number is 8 less than 13 less than 24 less than 7 less than 86?

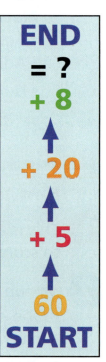

Now make up some number puzzles on your own. Test them out on a partner.

11. Use only addition and at least three comparisons.
12. Use only subtraction and at least three comparisons.
13. Use both addition and subtraction and at least three comparisons.

Chapter 15 Multiplication and Division

Dear Student,

In this chapter, you are going to go further with multiplication than you have ever gone before. You are going to look ahead to things you will learn about in fourth grade and beyond.

You will be multiplying and dividing larger numbers, and you will learn to answer questions like these:

Suppose there are 23 students in a class, and each student has read 48 books. How many books has the class read altogether?

Suppose there are 126 days until Dominick's birthday. How many weeks until his birthday?

We hope you enjoy this look into your mathematical future!

Mathematically yours,
The authors of *Think Math!*

Butterflies

Monarch

Would you like to be indoors and have beautiful butterflies circle around you and maybe even land on your hand? You can at an indoor butterfly garden. One such place is *Butterfly World* in Florida. There, you can see 50 different kinds of butterflies at any one time among the 5,000 that are on display.

Queen Alexandra's Birdwing

Orange Tiger

Blue Pansy

Solve.

1. Suppose you see 10 each of 9 different kinds of butterflies at *Butterfly World*. How many butterflies will you see in all?

2. Suppose you see 10 each of 24 different kinds of butterflies at *Butterfly World*. How many butterflies will you see in all?

3. *Butterfly World* provides guided tours. Suppose the tour guide gives 50 tours of 20 students each in a year. How many students does the tour guide lead in one year?

4. Adults can arrange to have children's birthday parties at *Butterfly World*. The cost is $18 per child with a minimum of 12 children. (That is, there must be at least 11 children.) Suppose it is your party. How many people do you want to have at your party? You may use the model at the right to help you find the total cost for that number of children.

Graphium Weiski

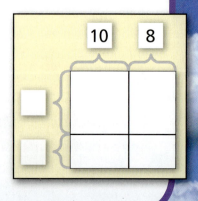

230 Chapter 15

An artist has collected different types of butterflies and put them in display cases. The table below shows the number of different butterflies in a case.

FACT·ACTIVITY 2

Use the information in the table for 1–3.

1. A museum store buys 25 displays of Birdwing butterflies. Write two related multiplication sentences to show the total number of butterflies in the display. Then write two related division sentences.

2. If the artist had 126 Zebra Swallowtail butterflies, how many cases could he make?

3. The artist has collected 44 Paris Peacock butterflies. How many cases can he make? Does he have any Paris Peacocks left over?

Butterfly Display Cases

Type of Butterfly	Number of Butterflies per Case
Birdwing	14
Paris Peacock	3
White Dragontail	8
Zebra Swallowtail	6

CHAPTER PROJECT

In the butterfly life cycle, caterpillars turn into butterflies. Make a caterpillar. You need 5 pompoms, 2 wiggly eyes, 1 pipe cleaner, and 1 clothespin for each caterpillar.

Directions: Cut the pipe cleaners into 1-inch-long pieces. Wrap 6 pieces of the pipe cleaner around the clothespin. Glue 5 pompoms together on top of the clothespin. Glue 1 pair of wiggly eyes on the face of the caterpillar.

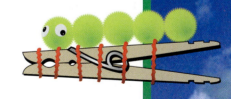

How many of each material do you need to make 25 caterpillars? How many caterpillars can you make with 1 pack of pompoms? 1 pack of wiggly eyes? 1 pack of pipe cleaners? 1 pack of clothespins? You may use sketches or base-ten blocks to help you.

Materials	Number per Pack
$\frac{1}{2}$- to 1-inch craft pompoms	300
wiggly eyes	152
12-inch long pipe cleaners	12
3-inch clothespins	36

Chapter 15 Lesson 1
EXPLORE
Dime Arrays

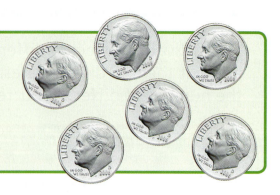

You want to make an array to help you find 8 × 60, but you do not have hundreds of counters. Think of a way to use dimes as counters so you can do this with fewer than 50 dimes.

1 Draw your array.

2 Complete the number sentence.

8 × 60 = ■

Now use fewer than 50 dimes to make an array that represents 6 × 80.

3 Draw your array.

4 Complete the number sentence.

6 × 80 = ■

Chapter 15, Lesson 3

REVIEW MODEL
Adding Partial Products

You can use a diagram and simpler problems to find a product.

Example Find 32 × 24.

	30	2
20	20 × 30 = **600**	20 × 2 = **40**
4	4 × 30 = **120**	4 × 2 = **8**

Each product in the diagram (600, 40, 120, and 8) is a **partial product**.

Think of the partial products as parts of the total product. They can be added together to find the total product.

You can record the partial products in a column to make it easier to add.

You can record the partial products in any order.

```
                    32
                 ×  24
        (20 × 30)  600
        (20 × 2)    40
        (4 × 30)   120
        (4 × 2)  +   8
                   768
```

Be sure to align the place values (ones, tens, and hundreds) to add.

✓ Check for Understanding

Use the diagram to find the partial products. Then add the partial products to find the total product.

1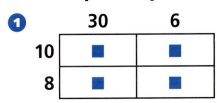

2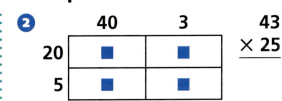

Chapter 15
Lesson 4

EXPLORE
Multiplying with Blocks and Money

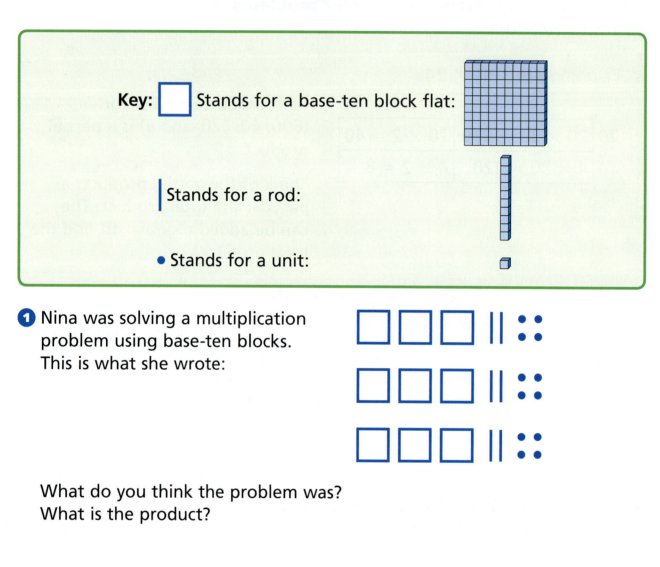

Key: ☐ Stands for a base-ten block flat:

| Stands for a rod:

• Stands for a unit:

1 Nina was solving a multiplication problem using base-ten blocks. This is what she wrote:

☐☐☐ || ::
☐☐☐ || ::
☐☐☐ || ::

What do you think the problem was?
What is the product?

2 Nina started to solve **243 × 3** by making **243** with dollars $1, dimes D, and pennies P.

$1 $1 D D D D P P P

How could Nina continue to solve **243 × 3**?

234 Chapter 15

Chapter 15
Lesson 6
EXPLORE
Finding Missing Streets

Trina counted 126 intersections on the 7 streets that run east to west on the map of her town.

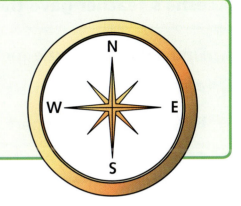

1. On a separate piece of paper, draw a map of the town.

2. How many streets on your map run north to south?

Mario counted 144 intersections on the 9 streets that run north to south on the map of his town.

3. Draw a map of the town on a separate piece of paper.

4. How many streets on your map run east to west?

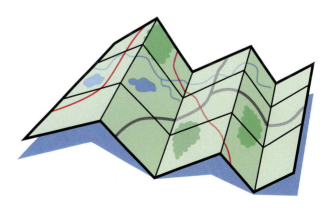

Chapter 15
Lesson 7
EXPLORE
A Division Puzzle Challenge

Keesha's teacher gave the class this challenge.

> Find a division puzzle for **144 ÷ 3** that splits **144** into two parts.
>
> **A** The first part must be a multiple of 10.
>
> **B** Make the second part as small as possible.
>
> **C** Use only whole numbers.

1 Keesha started with this puzzle.

$$\begin{array}{c} ? + ? \\ 3 \overline{)\, 140 + 4\,} \end{array}$$

Why won't her puzzle meet the challenge?

___

2 Keesha made this puzzle next.

$$\begin{array}{c} 30 + 18 \\ 3 \overline{)\, 90 + 54\,} \end{array}$$

Why won't this puzzle meet the challenge?

___

3 Find and complete the puzzle that meets the challenge.

Chapter 15 Lesson 8

REVIEW MODEL
Understanding Remainders

Sometimes you cannot divide objects evenly into groups.

Example Find 125 ÷ 4.

Step 1 Use base-ten blocks to represent the number you are dividing.

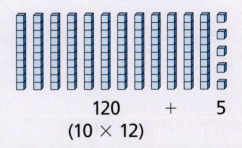

120 + 5
(10 × 12)

Step 2 Make 4 groups. Separate the tens into 4 equal groups by putting the same number of tens into each group.

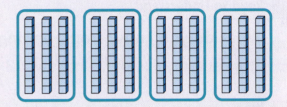

Step 3 Separate the ones into your 4 equal groups by putting the same number of ones into each group.

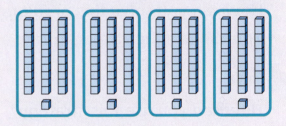

The **quotient** is 31—the number in each of the 4 groups.

You started with 5 ones. You only used 4 ones to make equal groups. So, you have 1 left over.

The **remainder** is 1—the number left over. So, 125 ÷ 4 = 31 r1.

✓ Check for Understanding

Use the picture to help you find the quotient and the remainder.

1 109 ÷ 5 = ■

2 368 ÷ 3 = ■

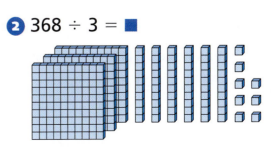

Chapter 15 Lesson 9

REVIEW MODEL
Problem Solving Strategy
Draw a Picture

> Janice's small dog plays in a rectangular grassy space in her backyard. The play space is 18 feet long and 15 feet wide. What is the area of the play space?

Strategy: Draw a Picture

Read to Understand

What do you know from reading the problem?

The rectangular play space is 18 feet long and 15 feet wide.

Plan

How can you solve this problem?

You can use the strategy *draw a picture.*

Solve

How can you use this strategy?

You can draw a picture of the play space. Draw a rectangle and label it with the length and width. Separate the rectangle into 4 parts and solve simpler problems by breaking the numbers into tens and ones. Find the area of each part, and add the 4 products to find the total area.

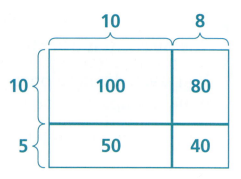

$10 \times 10 = 100$, $10 \times 8 = 80$, $5 \times 10 = 50$, and $5 \times 8 = 40$

$100 + 80 + 50 + 40 = 270$. So, the play space is 270 sq ft.

Check

Look back at the problem. Did you answer the question that was asked? Does the answer make sense?

Problem Solving Practice

Use the strategy *draw a picture*.

1. Jerry looked at a map of the town he was visiting and saw that the streets form 36 intersections. Some streets run east to west, and the other streets run north to south. How many streets could there be in the town?

2. Elena is decorating her bedroom. Her paint color choices are white, yellow, gray, or brown. She can choose red, green, blue, purple, orange, or pink curtains. How many choices does Elena have if she chooses one paint color and one curtain color?

Problem Solving Strategies

- ✔ Act It Out
- ✔ **Draw a Picture**
- ✔ Guess and Check
- ✔ Look for a Pattern
- ✔ Make a Graph
- ✔ Make a Model
- ✔ Make an Organized List
- ✔ Make a Table
- ✔ Solve a Simpler Problem
- ✔ Use Logical Reasoning
- ✔ Work Backward
- ✔ Write a Number Sentence

Mixed Strategy Practice

Use any strategy to solve. Explain.

3. Shakira has nickels, dimes, and quarters in her pocket. She pulls out 1 quarter. Then she pulls out 2 more coins. What are all the possible amounts of money Shakira could have pulled out of her pocket?

4. Tony and Fran were decorating cupcakes. Tony was faster than Fran. Every time Fran decorated one cupcake, Tony decorated two cupcakes. Together they decorated 48 cupcakes. How many cupcakes did each person decorate?

5. Becky drinks 6 glasses of water each day. She drinks 8 ounces each time she has a glass of water. How many ounces of water does Becky drink in a week?

6. Mr. Yang owns 4 pet stores. He ordered 216 bags of dog food. He wants each store to have the same number of bags of food. How many bags will each store receive?

Chapter 15 Vocabulary

Choose the best vocabulary term from Word List A for each sentence.

1. A(n) __?__ is an arrangement that shows objects in rows and columns.

2. A number that is multiplied by another number to find a product is called a(n) __?__.

3. A(n) __?__ of a whole number is a product of that number and another whole number.

4. The answer to a division problem is called the __?__.

5. A(n) __?__ is one of the set of numbers 0, 1, 2, 3, . . ., which continues without end.

6. The amount left over when a number cannot be divided equally is a(n) __?__.

7. The __?__ is the number to be divided in a division problem.

Word List A

array
column
diagram
dividend
divisor
factor
horizontal
intersecting lines
multiple
partial product
product
quotient
remainder
whole number

Complete each analogy using the best term from Word List B.

8. River is to horizontal as tree is to __?__.

9. Multiplication is to product as division is to __?__.

Word List B

factor
remainder
quotient
vertical

Talk Math

Discuss with a partner what you have learned about multiplication and division. Use the vocabulary terms *factor, multiple,* and *whole number.*

10. How can you use an array to multiply two numbers?

11. How does division relate to multiplication?

Venn Diagram

12 Create a Venn diagram for the words and terms related to multiplication and division. Label one circle *Multiplication* and the other circle *Division*.

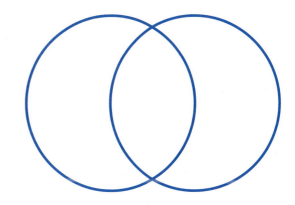

Word Definition Map

13 Create a word definition map for the term *array*.

A What is it?

B What is it like?

C What are some examples?

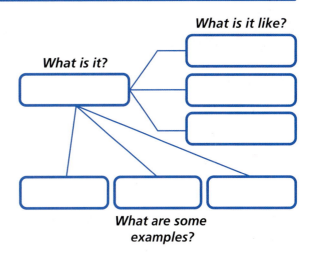

What's in a Word?

PRODUCT A *product* is a result. In math, a product is the result of multiplying two or more factors. In everyday life, there are other types of *products*. The *product* of a factory might be cars. The *product* of a farm might be corn. The *product* of a writer might be a book. The *product* of a musician might be a song. Each *product* is the result of performing an action.

GO ONLINE Technology
Multimedia Math Glossary
www.harcourtschool.com/thinkmath

GAME

Factor Factory

Game Purpose
To practice multiplying two-digit numbers using arrays

Materials
- 2 Number cubes
- 2 Colored pencils
- Centimeter grid paper, tape
- Calculator

How to Play the Game

1. This is a game for 2 players. Tape together 4 sheets of centimeter grid paper to make a 2-by-2 array. Choose your color of pencil.

2. To start, each player tosses one number cube. The two numbers tossed are the length and width of an array. The player who tossed the larger number draws the array on the grid paper and finds the number of squares in the array. That number is his or her score for the round. Remember to keep a record of your scores.

3. Now take turns. Toss one number cube. Decide whether to add that number of rows or columns to the array. Then figure out the new total number of squares. Your partner can use a calculator to check the answer. If it is correct, you get the number of squares that you added to the array as your points for the round.

4. Play until adding a row or column would make one side of the new array greater than 50. There is no score for that round. The game is over.

5. Add all your points. The player with more points wins.

Example:

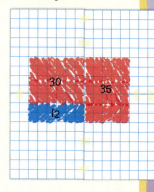

242 Chapter 15

GAME

Partial Claim

Game Purpose
To practice using partial products to find the product of 2 two-digit numbers

Materials
• Number cards (1–9, four sets)   • Calculator

How to Play the Game

1 This is a game for 2 players. Mix up the number cards. Place them in a stack face down.

2 Each player takes 2 cards to make a factor. The first number picked is the tens digit. The second number is the ones digit. Multiply the factors using partial products.
• Take turns. "Claim" (say aloud) and record the partial products of the two factors.
• Find the sum of your "claimed" partial products.
• Calculate and record the total product. Check your work with the calculator.

3 Play 6 rounds. Find the sum of your partial products from each round. The player with the higher total wins.

Here's a Sample Round

Suki picks 3 and 6 to form 36. Cal picks 1 and 7 to form 17. Suki and Cal multiply 36 × 17.

Suki claims the partial product 10 × 30 = 300. Cal claims 7 × 30 = 210. Suki claims 10 × 6 = 60. Cal claims 7 × 6 = 42.

Suki and Cal record the total product. They check the answer with a calculator.

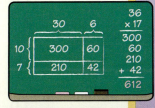

Chapter 15   243

CHALLENGE

Somebody erased some of the numbers from Trevor's math homework. Can you help Trevor find all the missing numbers?

For 1 to 4, find the missing factors.

①
```
    482
  ×   ■
   1200
    240
      6
  1,446
```

②
```
    264
  ×   ■
   1600
    480
     32
  2,112
```

③
```
    718
  ×   ■
   4200
     60
     48
  4,308
```

④
```
    963
  ×   ■
   4500
    300
     15
  4,815
```

For 5 to 8, find the missing partial products.

⑤
```
     39
  ×  35
    900
    270
    ■■■
     45
  1,365
```

⑥
```
     52
  ×  14
    ■■■
     20
    200
      8
    728
```

⑦
```
     47
  ×  63
   2400
    ■■■
    120
     21
  2,961
```

⑧
```
     26
  ×  37
    600
    180
    140
     ■■
    962
```

For 9 to 12, find all the missing digits.

⑨
```
     1■
  ×  57
    500
    ■■■
     70
     56
  1,026
```

⑩
```
     ■3
  ×  28
   ■■■■
     60
    480
     24
  1,764
```

⑪
```
     ■1
  ×  23
   1800
     20
    ■■■
      3
  2,093
```

⑫
```
     56
  ×  ■3
   ■■■■
    240
    150
     18
  2,408
```

Resources

Table of Measures 246

All the important measures used in this book are in this table. If you've forgotten exactly how many feet are in a mile, this table will help you.

Glossary 247

This glossary will help you speak and write the language of mathematics. Use the glossary to check the definitions of important terms.

Index 257

Use the index when you want to review a topic. It lists the page numbers where the topic is taught.

Table of Measures

METRIC | CUSTOMARY

LENGTH

METRIC	CUSTOMARY
1 decimeter (dm) = 10 centimeters	1 foot (ft) = 12 inches (in.)
1 meter (m) = 100 centimeters	1 yard (yd) = 3 feet, or 36 inches
1 meter (m) = 10 decimeters	1 mile (mi) = 1,760 yards, or 5,280 feet
1 kilometer (km) = 1,000 meters	

CAPACITY

METRIC	CUSTOMARY
1 liter (L) = 1,000 milliliters (mL)	1 cup (c) = 8 fluid ounces (fl oz)
	1 pint (pt) = 2 cups
	1 quart (qt) = 2 pints
	1 gallon (gal) = 4 quarts

MASS/WEIGHT

METRIC	CUSTOMARY
1 kilogram (kg) = 1,000 grams (g)	1 pound (lb) = 16 ounces (oz)
	1 ton (T) = 2,000 pounds

TIME

1 minute (min) = 60 seconds (sec)	1 year (yr) = 12 months (mo), or about 52 weeks
1 hour (hr) = 60 minutes	
1 day = 24 hours	1 year = 365 days
1 week (wk) = 7 days	1 leap year = 366 days

MONEY

1 penny = 1 cent (¢)
1 nickel = 5 cents
1 dime = 10 cents
1 quarter = 25 cents
1 half dollar = 50 cents
1 dollar ($) = 100 cents

SYMBOLS

<	is less than
>	is greater than
=	is equal to
≠	is not equal to
°F	degrees Fahrenheit
°C	degrees Celsius
(2,3)	ordered pair (x,y)

Glossary

PRONUNCIATION KEY

a	add, map	f	fit, half	n	nice, tin	p	pit, stop	yōō	fuse, few
ā	ace, rate	g	go, log	ng	ring, song	r	run, poor	v	vain, eve
â(r)	care, air	h	hope, hate	o	odd, hot	s	see, pass	w	win, away
ä	palm, father	i	it, give	ō	open, so	sh	sure, rush	y	yet, yearn
b	bat, rub	ī	ice, write	ô	order, jaw	t	talk, sit	z	zest, muse
ch	check, catch	j	joy, ledge	oi	oil, boy	th	thin, both	zh	vision,
d	dog, rod	k	cool, take	ou	pout, now	th	this, bathe		pleasure
e	end, pet	l	look, rule	o͝o	took, full	u	up, done		
ē	equal, tree	m	move, seem	o͞o	pool, food	û(r)	burn, term		

ə the schwa, an unstressed vowel representing the sound spelled *a* in above, *e* in sicken, *i* in possible, *o* in melon, *u* in circus

Other symbols:
• separates words into syllables
′ indicates stress on a syllable

A

acute angle An angle that has a measure less than a right angle (less than 90°)

Example:

addend [ad′end] Any of the numbers that are added

Example: 2 + 3 = 5

 addend addend

addition sentence [ə•dish′•ən sen′•təns] A number sentence that uses the operation of addition

algorithm [ăl′gə•rĭth′əm] A step-by-step method for solving a problem

amount [ə•mount′] The total number or quantity

area [âr′ē•ə] The number of square units needed to cover a flat surface

arrangement [ə•rānj′•ment] The way in which objects or numbers are grouped

array [ə•rā′] A rectangular arrangement of objects in rows and columns

Example:

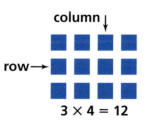

3 × 4 = 12

associative [ə•sō′•sē•tiv] Changing grouping

B

backward [bak′•wərd] The direction of a jump on a number line that represents subtraction

balance [bal′•əns] A tool used to weigh objects and to compare the weights of objects

bar graph [bär graf] A graph that uses bars to show data

Example:

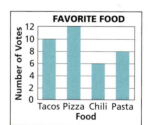

base-ten blocks [bās ten blăks] Blocks that model the base-ten system by representing ones, tens, and hundreds—units, rods of 10 units, flats of 10 rods, respectively

Example:

Glossary **247**

Glossary

capacity [kə·pa′sə·tē] The amount a container can hold

centimeter (cm) [sen′tə·mē·tər] A metric unit that is used to measure length or distance

cents [sents] A unit of money equal to one hundredth of a dollar

column [käl′əm] A vertical arrangement of objects

Example:

combination(s) [käm·bə·nā′·shən] A grouping of objects without regard to their order

combine [kəm·bīn′] To put together

commutative [kə·myōōt′·ə·tiv] The property of addition and multiplication that states that when the order of two or more addends or factors is changed, the sum or product is the same

compatible numbers [kəm·pat′·ə·bəl] Numbers that are easy to compute mentally

Example: Estimate. 528 ÷ 7

490 ÷ 7 = 70 ← 49 and 7 are compatible

560 ÷ 7 = 80 ← 56 and 7 are compatible

congruent figures [kən·grōō′·ənt] Figures that match exactly when one is placed on top of the other

Example:

coordinate [kō·ôrd′·n·it] The numbers in an ordered pair

Example:

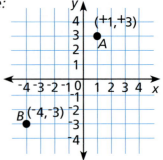

cost [kôst] The amount of money that must be paid to purchase an item

could [kəd] May be

Cross Number Puzzle [krôs num′·bər puz′·əl] A way of modeling addition or subtraction so that the values in each place are computed separately before the final answer is found

cube [kyōōb] A solid figure with six congruent square faces

Example:

cubic centimeter [kyōō′·bik sen′·tə·mēt′·ər] A unit for measuring volume where the base unit is a cube that is one centimeter in length, width, and height

cubic unit [kyōō′·bik yōō′·nət] A cube with a side length of one unit; used to measure volume

cup (c) [kup] A customary unit used to measure capacity

248 Glossary

Multimedia Math Glossary   www.harcourtschool.com/thinkmath

Glossary

data [dā′tə] Information from which conclusions can be made

decrease [dē′•krēs] To make smaller

denominator [di•nä′mə•nā•tər] The number in a fraction below the line, which tells how many equal parts there are in the whole or in the group

Example: $\frac{3}{4}$ ← denominator

diagonal [di•ag′•ə•nəl] A line that connects two opposite corners of a figure

diagram [di′•ə•gram] A drawing that can be used to represent a situation

difference [dif′rən(t)s] The result of subtraction

Example: $6 - 4 = 2$
　　　　　　　└difference

digits [di′jəts] One of the symbols 0, 1, 2, 3, 4, 5, 6, 7, 8, 9, or 10 used in a written representation of a number

dividend [di′və•dend] The number that is to be divided in a division problem

Example: $35 \div 5 = 7$
　　　　　　└dividend

division sentence [də•vi′•zhən sen′təns] A number sentence that uses the operation of division

Example: $21 \div 3 = 7$

divisor [di•vi′zər] The number that divides the dividend

Example: $35 \div 5 = 7$
　　　　　　　　└divisor

dozen [duz′•ən] A set of 12

Example: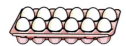

earlier [ər′•lē•ər] A time that occurs before another time

east [ēst] On a compass rose, the direction that horizontally points to the right

edge [ej] A line segment formed where two faces meet

Example: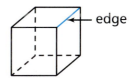

elapsed time [i•lapst′ tīm] The amount of time that passes from the start of an activity to the end of that activity

eliminate [ē•lim′•ə•nāt] To narrow down

equal [ē′•kwəl] Having the same value

equal (=) [ē′•kwəl] A symbol used to show that two numbers have the same value

Example: $384 = 384$

equally likely as [ē′•kwəl•lē li′klē as] Having the same chance of happening

equivalent [ē•kwiv′ə•lənt] Two or more sets that name the same amount

equivalent fractions [ē•kwiv′ə•lənt frak′shənz] Two or more fractions that name the same amount

Example: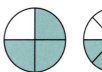

$\frac{3}{4} = \frac{6}{8}$

estimate [es′tə•māt] *verb:* To find about how many or how much

estimate [es′tə•mit] *noun:* A number close to an exact amount

even [ē′vən] A whole number that has a 0, 2, 4, 6, or 8 in the ones place

Glossary **249**

Glossary

exchange [eks•chānj′] To trade things of equal value

expanded form [ik•spand′id fôrm] A way to write a number by showing the value of each digit

Example: 7,201 = 7,000 + 200 + 1

experiment [ik•sper′ə•mənt] An exploration of a probability question, such as, "What's the likelihood of getting heads in a coin toss?"

expression [ik•spre′shən] The part of a number sentence that combines numbers and operation signs, but doesnt have an equal sign

Example: 5 × 6

face [fās] A flat surface of a solid figure

Example:

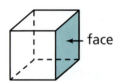

fact family [fakt fam′ə•lē] A set of related addition and subtraction, or multiplication and division, number sentences

Example: 4 × 7 = 28   28 ÷ 7 = 4
7 × 4 = 28   28 ÷ 4 = 7

factor [fak′tər] A number that is multiplied by another number to find a product

Example: 4 × 7 = 28   4
                          ×7
                          ──
                          28

The factors are 4 and 7.

Fahrenheit [fer′ən•hīt] A customary unit for measuring temperature

Example:

Fahrenheit thermometer

fewest [fyoo′•əst] Less in number than all other groups

fewest units [fyoo′•əst yoo′•nit] The least possible units

flat [flat] The base-ten block that represents 10 rods or 100 units

flip [flip] A movement of a figure to a new position by flipping the figure over a line

Example:

fluid ounce [ouns] A unit of liquid capacity

forward [fôr′wərd] The direction of a jump on a number line that represents addition

fourth [fôrth] The term to describe each of four fractional parts

Example:

fourths [fôrths] When something is divided into four fractional parts

fraction [frak′shən] A number that names part of a whole or part of a group

Example: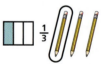

gallon (gal) [ga′lən] A customary unit for measuring capacity

Example: 1 gallon = 4 quarts

graph [graf] A way to organize data and information

greater than (>) [grā′tər than] A symbol used to compare two numbers, with the greater number given first

Example: 6 > 4

greatest [grā′test] More in value than all other values

grid [grid] Horizontal and vertical lines on a map

group [groop] To put things together

Glossary

half [haf] One part of something when it is divided into two equal parts

half inch [haf inch] One part of an inch when the inch is divided into two equal parts

halves [havz] When something is divided into two equal fractional parts

Example:

height [hīt] The distance from the bottom to the top of something standing upright

horizontal [hôr′•ə•zon′•təl] The direction from left to right

horizontal axis [hôr′•ə•zon′•təl] The horizontal number line on a coordinate plane

hour (hr) [our] A unit used to measure time; in one hour, the hour hand on a clock moves from one number to the next; 1 hour = 60 minutes

hundreds [hun′drəds] The place value that represents 100 through 900

impossible [im•pä′sə•bəl] An event that will never happen

increase [in′•krēs] To make larger

input [in′•pût] The number that is put into an input-output table or algebraic equation

intersecting lines [in•tər•sek′ting linz] Lines that cross

Example:

intersection [in•tər•sek′•shən] The place where two streets or lines cross each other

label [lā′•bəl] A term used in a graph to classify data

later [lāt′•ər] A time that occurs after another time

least [lēst] Less in value than all other values

length [lenkth] The measure of a side of a figure

less likely than [les lik′•lē than] Not as likely to happen than something else

less than (<) [les than] A symbol used to compare two numbers, with the lesser number given first

Example: 3 < 7

line symmetry [līn si′mə•trē] If one can fold a figure so the two parts match exactly, then the fold line is called a line of symmetry for that figure

magic square [maj′•ik skwer] A square array of numbers where every row, column, and diagonal has the same sum

millions [mil′yəns] The place value to the left of hundred thousands

minute (min) [min′it] A unit used to measure short amounts of time; in one minute, the minute hand moves from one mark to the next

more likely than [môr lik′•lē than] A greater chance of happening than something else

most [mōst] More in number than all other groups

multi-digit number [mul′•tē dij′•it num′•bər] A number that has two or more digits

Glossary **251**

Glossary

multiple [mul′tə·pəl] A number that is the product of a given number and another whole number

Example:
```
  10   10   10   10
  ×1   ×2   ×3   ×4
  10   20   30   40  ← multiples of 10
```

must [must] Has to be

negative number [neg′·ə·tiv num′·bər] A number that lies to the left of zero on the number line or below zero on a thermometer

Example:

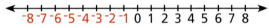

The red numbers are negative numbers.

negative sign (−) [neg′·ə·tiv sin] A symbol used to show a negative number

net [net] A two-dimensional pattern that can be folded to make a three-dimensional figure

Example:

north [nôrth] On a compass rose, the direction that vertically points upward

number [num′·bər] A value composed of one or more digits

number line [num′·bər lin] A line with equally spaced tick marks named by numbers

Example: 0 1 2 3 4 5 6 7 8

number sentence [num′·bər sen′təns] A statement that shows the relationship between two or more values

Example: 5 + 3 = 8 is a number sentence.

numerator [noo′·mə·rā·tər] The number in a fraction above the line, which tells how many parts are being counted

Example: $\frac{3}{4}$ ← numerator

odd [od] A whole number that has a 1, 3, 5, 7, or 9 in the ones place

ones [wuns] The place value that represents the numbers 1 through 9

operation sign [äp·ə·ra′·shen] A symbol which represents the process that will change one number to another according to a rule

Example: operations signs: +, −, ÷, ×

ounce (oz) [ouns] A customary unit for measuring weight

outcome [out′kum] A possible result of an experiment

output [out′püt] The number that is the outcome of an input-output table or an algebraic equation

package [pak′·ij] To group together

parallel [par′·ə·lel] Lines that never cross, lines that are always the same distance apart

Example:

parallelogram [par·ə·lel′ə·gram] A quadrilateral with 2 pairs of parallel sides and 2 pairs of equal sides

Example:

part [pärt] One portion of a whole

Example: A quarter hour is $\frac{1}{4}$ of an hour, and an hour has 4 such parts in it.

partial product [pär′·shəl prä′dəkt] A part of a final product that is the result of multiplying the ones, tens, hundreds, and so on, separately

Example:
```
     24
   ×  3
     12 ← Multiply the ones: 3 × 4 = 12
   +60 ← Multiply the tens: 3 × 20 = 60
     72
```

252 Glossary

Glossary

pattern [pat′ərn] A rule that allows you to predict what is missing or what comes next in a sequence of numbers or objects; casually, the word pattern can be used for the sequence itself

Examples: 2, 4, 6, 8, 10

pentagon [pen′tə•gän] A polygon with five sides

Example:

perimeter [pə•ri′mə•tər] The distance around a figure

Example:

perpendicular [pər•pən•di′kyə•lər] Lines that form four equal angles

Example:

pictograph [pik′tə•graf] A graph that uses pictures to show and compare information

Example:

piece [pēs] A part of a whole

pint (pt) [pīnt] A customary unit for measuring capacity

Example: 1 pint = 2 cups

polygon [pol′ē•gän] A closed plane figure with straight sides that are line segments

Example:

possible [pos′ə•bəl] Something that has a chance of happening

pound (lb) [pound] A customary unit for measuring weight

Example: 1 pound = 16 ounces

predict [pri•dikt′] To make a reasonable guess about what will happen

prediction [prē•dik′shən] A reasonable guess about what will happen

price [prīs] The amount of money that a customer will be charged to buy something

prism [priz′əm] A solid figure that has two congruent, polygon-shaped bases, and other faces that are rectangles

Example:

rectangular prism   triangular prism

probability [prä•bə•bil′ə•tē] The study of random occurences

product [prä′dəkt] The result of multiplication

Example: 3 × 8 = 24
              ↳ product

purchase [pər′•chəs] To get by paying money for

pyramid [pir′ə•mid] A solid, pointed figure with a flat base that is a polygon

Example:

Q

quadrilateral [kwä•drə•la′tə•rəl] A polygon with four sides

Example:

quart (qt) [kwôrt] A customary unit for measuring capacity

Example: 1 quart = 2 pints

quarter [kwôrt′•ər] One fourth of something

quarter after [kwôrt′•ər af′•tər] One fourth of an hour, or 15 minutes, after the hour

quarter inch [kwôrt′•ər inch] One fourth of an inch

Glossary

quarter past [kwôrt′•ər past] One fourth of an hour, or 15 minutes, after the hour

quotient [kwō′shənt] The number, not including the remainder, that results from division

Example: 8 ÷ 4 = 2
             ↳ quotient

range [rānj] The difference between the largest number and the smallest number in a set of data

rectangle [rek′tang•əl] A quadrilateral with 2 pairs of parallel sides, 2 pairs of equal sides, and 4 right angles

Example:

rectangular prism [rek•tan′gyə•lər pri′zəm] A solid figure with six faces that are all rectangles

Example:

reflection [rē•flekt′shən] A transformation that creates a mirror image of an object

Example:

region [rē′•jən] A section or part of a larger whole

regroup [rē•grōōp′] To exchange amounts of equal value when working with a number

Example: 5 + 8 = 13 ones or 1 ten 3 ones

remainder [ri•mān′dər] The amount left over when a number cannot be divided evenly

right angle [rīt ang′gəl] A special angle that forms a square corner; a right angle measures 90°

Example:

rod [räd] A base-ten block that represents one row of 10 units

Example:

round [roun′d] To replace a number with another number close to the same value

row [rō] A horizontal arrangement of objects

Example: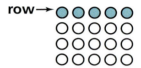

rule [rōōl] A procedure for solving a mathematical problem

scale [skāl] The numbers on a bar graph that help you read the number each bar shows

section [sek′•shən] A part or piece of a whole

separate [sep′•ə•rāt] To pull apart

separation [sep•ə•rā′•shən] The act of pulling something apart, such as partitioning a whole number into a sum of smaller whole numbers

shorthand notation [shôrt′•hand nō•tā′•shən] Mathematical shorthand used to express numerical computations

skip-count [skip′•kount] A method of counting where certain numbers are skipped so that each space between the numbers has the same value

slide [slīd] A movement of a figure to a new position without turning or flipping it

Example: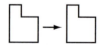

254 Glossary   Multimedia Math Glossary   www.harcourtschool.com/thinkmath

Glossary

smallest unit [smôl′·əst yōō′·nit] The littlest unit possible

south [south] On a compass rose, the direction that vertically points downward

spaces [spās′·əs] The parts of a number line that are between the tick marks

spend [spend] The amount of money used to purchase items

square [skwâr] A quadrilateral with 2 pairs of parallel sides, 4 equal sides, and 4 right angles

Example:

standard algorithm [stan′·dərd al′·gə·rith·əm] A step-by-step procedure for computing that will work for every event within a certain type of number set

subtraction sentence [səb·trak′·shən] A number sentence that uses the operation of subtraction

sum [sum] The result of addition

survey [sər′vā] A method of gathering information

symbol [sim′·bəl] Something that stands for something else, such as a picture symbol in a pictograph

symmetry [sim′ə·trē] A figure has symmetry if it can be folded along a line so that the two parts match exactly; one half of the figure looks like the mirror image of the other half

table [tā′·bəl] A tool used to record data

Example:

Favorite Sport	
Sport	Number
Soccer	||||| |||
Baseball	|||
Football	|||||
Basketball	|||||	

temperature [tem′·pər·ə·chər] How hot or cold something is

tens [tens] The place value that represents 10 through 90

thermometer [thər·mäm′·ət·ər] A tool used to measure temperature

thirds [thərds] When something is divided into three equal parts

thousand [thou′·zənd] How the thousands place-value is referred to when reading a number out loud

thousands [thou′·zənds] The place value that represents 1,000 through 9,000

three-dimensional [thrē də·men′·chə·nəl] A figure that has length, width, and height

times (×) [tīmz] The operation of multiplication

ton [tun] A customary unit of measuring large weights, equal to 2,000 pounds

total [tōt′·təl] Another word for sum

trade [trād] To exchange for something of equal value

trapezoid [tra′pə·zoid] A quadrilateral with one pair of parallel sides

Example:

trend [trend] On a graph, the area where data increase, decrease, or stay the same over time

triangle [trī′ang·əl] A polygon with three sides

Examples: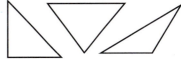

triangular prism [trī′an′·gyə·lər priz′əm] A solid figure that has two parallel faces that are triangles and three faces that are rectangles

turn [tûrn] A movement of a figure to a new position by rotating the figure around a point

Example:

Glossary

ungroup [un•groop′] To take apart amounts of equal value to rename a number

unit [yoo′nĭt] A base-ten block that represents 1 (small cube)

unpackage [un•pak′ĭj] To ungroup

value [val′•yoo] What something is worth

variable [vâr′ē•ə•bəl] A symbol or letter that stands for an unknown number

vertex [vûr′teks] The point at which two or more line segments meet in a plane figure or where three or more edges meet in a solid figure

Example:

vertex

vertical [vər′•tĭ•kəl] The direction from top to bottom

vertical axis [vər′•tĭ•kəl ak′•səs] The vertical number line on a coordinate plane

Example:

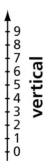

volume [väl′yəm] The amount of space a solid figure takes up

west [west] On a compass rose, the direction that horizontally points to the left

Example:

whole number [hōl num′bər] One of the numbers 0, 1, 2, 3, 4 . . . ; the set of whole numbers goes on without end

width [width] A measurement of a figure from one side to another

Index

A

Act It Out, 10–11, 206–207

Addends
find missing, in magic square, 39

Addition
adding partial products, 233
coins, 54
Cross Number Puzzle, 77
estimate, 73, 74
    closest multiple of ten, 43
    rounding, 43
expanded form to add, 217, 220
on grid, 93
multi-digit numbers, 216, 221
number line to, 7
regrouping, 75
Student Letter, 35, 67, 213

Addition puzzles, 5

Addition sentences
completing, 9

Algorithms, 67

Angles
right, 173

Area, 151
measuring, 155–157
Student Letter, 151

Arrays
dime array, 232
rectangles to represent, 189
separating arrays to make simpler problems, 143
separating to multiply, 190

B

Base-ten blocks
addition with, 75
comparing numbers, 70
estimating addition, 73
multiplying with, 187, 188, 234
regrouping, 55

C

Capacity
compare, 205
Student Letter, 199

Centimeter
area, 155–157
perimeter, 155–157

Challenge, 16, 34, 50, 66, 84, 102, 118, 134, 150, 166, 182, 198, 212, 228, 244

Charts, 119
place value, 59
Student Letter, 119

Classify
events as possible or impossible, 123–124

Class Store
price at, 126

Coins
adding, 54
dime array, 232
fewest coins, 41
grouping, 54
multiplying money, 186
multiplying with, 234
subtracting, 54

Combinations
listing, 25

Compare
capacity, 205
fractions, 111
place value, 70
three-dimensional figures, 175

Counting
shortcuts, 4
strategies for, 4

Cross Number Puzzle
adding, 77
subtracting, 77

Index

D

Data
  graph to find rule, 88
  pictographs, 122

Dime array, 232

Distance
  time and speed, 203

Division
  division puzzle challenge, 236
  remainders, 237
  Sharing Machine, 94
  solve division problems, 191
  Student Letter, 229

Division fact family
  writing, 141

Division sentences
  writing, 95

Dozen
  fractions of, 108–109

Draw a Picture, 160–161, 238–239

E

Equivalent fractions, 107

Estimation
  addition, 73, 74
    closest multiple of ten, 43
    rounding, 43
  eliminating possibilities, 42
  subtraction, 76
    closest multiples of ten, 43
    rounding, 43
  weight, 204

Even numbers
  exploring, 40

Expanded form
  to add, 217, 220
  to subtract, 219–220

F

Factors
  intersecting lines to find, 27
  missing, 140
  using ten as, 142

Figures
  parallel sides, 170, 173
  pentagons, 173
  polygons, 172–174
  prisms, 175
  pyramids, 175
  quadrilateral, 172, 173
  rectangle, 172
  three-dimensional figures, 174–175
  triangle, 173

Find a Rule cards, 89

Fourths
  dozen, 109
  equivalent fractions, 110
  hour, 110
  identify, 106
  inch, 154

Fractions, 103
  compare with models, 111
  of dozen, 108–109
  equivalent, 107
  of hour, 110
  names for, 106
  Student Letter, 103

G

Games
  Addition Scramble, 227
  Caught in the Middle, 149
  Factor Factory, 242
  Factor Maze, 33
  Factor Tic-Tac-Toe, 197
  Find a Rule, 100
  Fit!, 32
  Fraction Construction Zone, 116
  The Freezing Game, 210
  Least to Greatest, 49, 83
  Make a Rule, 101

258 Index

Index

Marble Mystery, 117
Missing Operation Signs, 15
Multiplication Challenge, 196
Number Line Grab, 14
Ordering Numbers, 82
Partial Claim, 243
Perimeter Golf, 165
Place Value Game, 65, 226
Polygon Bingo, 181
Ruler Game, 164
Tic-Tac-Toe Multiplication, 148
Time Concentration, 211
Trading to 1,000, 64
What Are My Coins?, 48
What's My Rule?, 180
Where's My Car?, 132
Where's My House?, 133

Geometry, 167
exploring polygons, 173
identifying parallel lines, 171
parallel sides of quadrilaterals, 170
sorting polygons, 173
Student Letter, 167
three-dimensional figures, 174–175

Graphs, 119
graph to find rule, 88
pictographs, 122
Student Letter, 119

Grids
adding on, 93
finding missing streets, 235
map grid, 127
multiplication, 138–139
subtracting on, 93

Grouping
base-ten, 70
coins, 54
Student Letter, 51

Guess and Check, 144–145

H

Half
dozen, 108
hour, 110

identify, 106
inch, 154

Hour
fractions of, 110

I

Impossible events, 123–124

Inch
measuring, 154

Intersecting lines
to find factors, 27
multiplication, 138–139, 188

Intersections
exploring, 20, 21
hidden, 22
missing, 22
pairing objects, 23, 24

L

Length, 151
measure to nearest inch, one half inch, and one fourth inch, 154
Student Letter, 151

Less likely, 124

Line of symmetry, 173

Lines
exploring, 20
identify parallel lines, 171
intersecting, 20, 21

Look for a Pattern, 96–97, 176–177

M

Magic square
introducing, 38
missing addends, 39

Make a Model, 112–113

Make an Organized List, 60–61

Index

Make a Table, 128–129

Map grid
    finding missing streets, 235
    name locations on, 127

Measurement
    area, 155–157
    capacity, 205
    length, 154
    perimeter, 155–157
    time, 202
    volume, 158–159
    weight, 204

Money
    multiplying, 186, 234
    price at class store, 126

More likely, 124

Multi-digit numbers
    addition, 216, 221
    subtraction, 218

Multiples
    of ten
        estimate addition, 43
        estimate subtraction, 43

Multiplication, 135
    adding products of simpler multiplication problem, 188
    base-ten blocks, 187, 188
    with base-ten blocks, 234
    counters, 138–139
    dime array, 232
    grid paper, 138–139
    intersecting lines, 138–139, 188
    missing factors, 140
    models to help, 138–139
    money, 186
    with money, 234
    practice, 138–139
    products, 138
    rectangles to represent arrays, 189
    separating arrays, 143, 190
    square tiles, 138–139
    strategies for, 183

Student Letter, 17, 135, 183, 229
using ten as factor, 142

Multiplication fact family
    write, 141

Multiplication sentences
    pairing objects, 25
    write, 26, 140, 186

Mystery Number Puzzle
    order of clues, 58
    place value, 57
    possible numbers, 56

N

Number line
    to add, 7
    jumps on, 71
    locating six on, 6
    to subtract, 7, 72

Number Line Hotel, 92–93

Number patterns, 8

Numbers
    comparing with place value, 70

Number sentences
    completing, 9

O

Odd numbers
    exploring, 40

Operations
    Student Letter, 1

Order
    clues in Mystery Number Puzzle, 58

P

Pairs
    listing combinations, 25
    of objects, 22, 24

Index

Parallel lines
    identify, 171

Parallel sides
    of quadrilateral, 170
    sorting polygons, 173

Partial products
    adding, 233

Patterns
    find a rule for, 91
    look for, 90
    multiple of ten, 142
    number, 8
    Number Line Hotel, 92
    rules and, 85
    Student Letter, 85

Pentagons
    sorting, 173

Perimeter
    measuring, 155–157

Pictographs, 122

Place value
    chart for, 59
    Mystery Number Puzzle, 57
    Student Letter, 51

Polygons
    explore, 172
    sorting, 173

Possibilities
    eliminating, 42

Possible events, 123–124

Price, 126

Prisms, 175

Probability
    classify events as possible or impossible, 123–124
    describing likelihood of event, 124
    listing outcomes, 125

Problem Solving Strategies
    Act It Out, 10–11, 206–207
    Draw a Picture, 28–29, 160–161, 238–239
    Guess and Check, 144–145
    Look for a Pattern, 96–97, 176–177
    Make a Model, 112–113
    Make an Organized List, 60–61
    Make a Table, 128–129
    Solve a Simpler Problem, 78–79, 222–223
    Work Backward, 44–45, 192–193

Products, 138
    adding partial, 233

Pyramids, 175

Q

Quadrilateral
    parallel sides, 170
    sorting, 173

R

Rectangle, 172

Regrouping
    addition with, 75
    base-ten blocks, 55
    Student Letter, 51

Remainders, 237

Right angle, 173

Rounding
    to estimate addition, 43

Rules, 85
    Find a Rule cards, 89
    find rule for pattern, 91
    graph to find, 88
    Student Letter, 85

S

Sharing Machine, 94

Six
    locating on number line, 6

Solve a Simpler Problem, 78–79, 222–223

Index

Speed
    time and distance, 203

Square tiles
    multiplication, 138–139

Student Letter
    addition and subtraction in depth, 213
    building operations, 1
    charts and graphs, 119
    exploring multiplication, 135
    fractions, 103
    geometry, 167
    grouping, regrouping, and place value, 51
    length, area, and volume, 151
    multiplication, 135
    multiplication and division, 229
    multiplication situations, 17
    multiplication strategies, 183
    rules and patterns, 85
    time, temperature, weight, and capacity, 199
    understanding addition and subtraction algorithms, 67
    using addition and subtraction, 35

Subtraction
    coins, 54
    Cross Number Puzzle, 77
    estimate, 76
        closest multiples of ten, 43
        rounding, 43
    expanded form to subtract, 219, 220
    on grid, 93
    multi-digit numbers, 218
    number line, 7, 72
    Student Letter, 35, 67, 213

Subtraction sentences
    completing, 9

Symmetry, line of, 173

T

Temperature
    Student Letter, 199

Tens
    dime array, 232
    as factor, 142
    multiples of
        estimate addition, 43
        estimate subtraction, 43

Thirds
    dozen, 109
    equivalent fractions, 110
    identify, 106

Three-dimensional figures, 174–175

Time
    distance and speed, 203
    reading, 202
    Student Letter, 199

Triangle
    right angles, 173

V

Vocabulary, 12–13, 30–31, 46–47, 62–63, 80–81, 98–99, 114–115, 130–131, 146–147, 162–163, 178–179, 194–195, 208–209, 224–225, 240–241

Volume, 151
    measurement, 158–159
    Student Letter, 151

W

Weight
    estimate, 204
    Student Letter, 199

Wholes
    parts of, 106

Work Backward, 44–45, 192–193

World Almanac for Kids, 2–3, 18–19, 36–37, 52–53, 68–69, 86–87, 104–105, 120–121, 136–137, 152–153, 168–169, 184–185, 200–201, 214–215, 230–231

Photo Credits

Page Placement Key: (t) top, (b) bottom, (c) center, (l) left, (r) right, (bg) background, (i) insert.

CHAPTER 1: 2 (tcl) Corel; (bcl) Corel; (tcr) Corel; (r) Ingram Publishing; (bg) Corel; 3 (tr) Jonathan Knowles/Masterfile.

CHAPTER 2: 18 (bg) Corel; (b) Paul Thompson; Ecoscene/CORBIS.

CHAPTER 3: 36 (bg) Maximilian Stock Ltd./Foodpix; (tr) Tim Laman/National Geographic Image Collection; (cl) Natural Visions/Alamy; (br) Otto Stadler/Peter Arnold; 37 (t) David Loftus/Getty Images.

CHAPTER 4: 52 (bg) Daryl Benson/Masterfile; (b) Daniel J. Cox/CORBIS; 53 (tr) Jim West/Alamy; (tl) Atlantide Phototravel/CORBIS; (bcr) Masterfile; (c) Daryl Benson/Masterfile.

CHAPTER 5: 68 (tr) CORBIS; (bg) Shannon Fagan/Getty Images; (br) iStockphoto; 69 (br) Holger Winkler/zefa/CORBIS.

CHAPTER 6: 86 (bg) Harry Sieplinga/Getty Images; (tr) Alamy; (c) Steve Gorton/Getty Images; 87 (r) Ousekh necklace, New Kingdom (faience), Egyptian, 18th Dynasty (c. 1567–1320 BC)/Private Collection/The Bridgeman Art Library International; (cl) Allen Rockwell/Alamy.

CHAPTER 7: 104 (cl) Scott B. Rosen/Bill Smith Studio; (c) Age Fotostock; (cr) Simon de Glanville/Alamy; (tr) SuperStock; (bg) Jupiter Images; 105 (l) Melissa Lockhart/SuperStock; (cl) Garry Gay/Alamy; (cr) Alamy; (r) Alamy; (tr) Melissa Lockhart/SuperStock.

CHAPTER 8: 120 (bg) Quiksilver/DC/NewsCom; (tr) Joe McBride/Getty Images; 121 (r) China Newsphoto/Reuters/CORBIS.

CHAPTER 9: 136 (t) PictureArts/NewsCom; (c) Jupiter Images; (tr) PictureArts/NewsCom; (bl) Scott B. Rosen/Bill Smith Studio; (br) Scott B. Rosen/Bill Smith Studio; 137 (tl) Scott B. Rosen/Bill Smith Studio; (tc) Scott B. Rosen/Bill Smith Studio; (tc) David Young-Wolff/PhotoEdit; (tr) Darin Burt Photography, photographersdirect; (cl) Brad & Melody Niese Photography, photographersdirect; (cr) Paolo Aristide Carboni Photography, photographersdirect; (r) Scott B. Rosen/Bill Smith Studio; (bl) Scott B. Rosen/Bill Smith Studio.

CHAPTER 10: 152 (bg) Frank Rothe/Getty Images; (cl) BIOS Gilson François/Peter Arnold; (c) Stephen Frink/CORBIS; (cr) BIOS Gilson François/Peter Arnold; 153 (cr) PHONE Labat Jean-Michel/Peter Arnold.

CHAPTER 11: 168 (b) Asa Gauen/Alamy; (bg) Richard T. Nowitz/CORBIS; 169 (cl) SuperStock/Age Fotostock; (c) Jupiter Images; (r) Alamy

CHAPTER 12: 184 (b) AP Photo/U.S. Postal Service; (br) AP Photo/Nick Ut; (tl) Smithsonian National Postal Museum; (tc) Smithsonian National Postal Museum; (tc) Tom Pantages; (tr) ZUMA Archive/NewsCom; 185 (tc) Alamy; (tr) Richard Naude/Alamy; (cr) Liquid Light/Alamy; (c) Gary W. Sargent; (c) Raimist Photography, photographersdirect

CHAPTER 13: 200 (bg) CORBIS; (tc) Alamy; (cl) Getty Images; (bl) Alaska Stock LLC/Alamy; (tl) Alamy; (cr) Getty Images; 201 (br) Roger Moen; (tc) Jupiter Images; (tr) Jupiter Images

CHAPTER 14: 214 (bg) Alamy; 215 (t) Robbie Rosenfeld

CHAPTER 15: 230 (bg) Alamy; (tc) Bristol City Museum/Nature Picture Library; (tr) Alamy; (c) iStockphoto; (cr) Robert & Lorri Franz/CORBIS; (br) Alamy; 231 (tr) Alamy

All other photos © Harcourt School Publishers. Harcourt photos provided by the Harcourt Index, Harcourt IPR, and Harcourt photographers: Weronica Ankarorn, Eric Camden, Doug Dukane, Ken Kinzie, April Riehm, and Steve Williams.